Compact Power for Pulsed Applications: A Guide to Capacitor Charger Design

Molina

First Printing, 2024

TABLE OF CONTENTS

CHAPTER

I. INTRODUCTION 1

1.1. Background & Application 1
1.1.1. Pulsed Power 1
1.1.2. Applications of Pulsed Power 2
1.1.3. Capacitor Charger Application 4
1.2. Motivation 4
1.2.1. System Architecture 5
1.2.2. Power Topology 6
1.2.3. Modularity 7
1.2.4. Semiconductor Material 9
1.3. book Organization 11

II. LITERATURE REVIEW 13

2.1 IPOS Converters 13
2.2 Pulsed Power Capacitor Chargers 16
2.3 IPOS Pulsed Power Capacitor Chargers 20
2.4 Summary 24

III. DESIGN & SIMULATION 26

3.1 Overview 26
3.2 Mathematical Analysis 26
3.3 Part Selections 29
3.3.1 FET's 29
3.3.2 Gate Driver 30
3.3.3 Input Filter 33
3.3.4 Transformer 35

3.3.5 Rectifier Diodes 39
3.3.6 Output Inductor 41
3.4 System Simulation 42

IV. EXPERIMENTAL EVALUATION 52

4.1 Charger Module Board 52
4.2 Test Setup 54
4.2.1 Input Test CCA 56
4.2.2 Output Test CCA 57
4.2.3 FPGA 59
4.2.4 Modular Stack Setup 61
4.3 Results 63
4.3.1 Single Card Testing 63
4.3.2 Four Card Testing 69

V. CONCLUSION & FUTURE WORK 76

5.1 Conclusion 76
5.2 Future Work 77

REFERENCES 79

CHAPTER I

INTRODUCTION

1.1. Background & Application

1.1.1. Pulsed Power

Pulsed power is a rapidly growing field in the electrical engineering community. It primarily focuses on the delivery of energy in a small amount of time. In his definitive textbook on the subject, J.C. Martin defines pulsed power as the "storage of electrical energy over relatively long times and then its rapid release over a comparatively short period" [1]. The small time transients in pulsed power make it unique from other conventional fields of electrical engineering dealing with high power. In the power generation and transmission fields, the delivered energy, while substantial, is consumed by the load over the span of seconds, minutes, or even hours. Pulsed power applications often deliver their energy to the load in the span of milliseconds, even ranging down to picoseconds. To illustrate the physics at play in pulse power, the generally acknowledged equations for potential energy (E), power (P), and charge (Q) of a system are given below:

$$E = Q * V \tag{1.1}$$

$$P = V * I \tag{1.2}$$

$$Q = I * t \tag{1.3}$$

When the equations for power and charge (Equations 1.2 and 1.3) are rearranged and substituted into Equation 1.1, the following equation for energy can be extrapolated:

$$E = P * t \tag{1.4}$$

Equation 1.4 emphasizes the driving principle behind pulsed power, that energy is directly related to power and time (*t*). If the energy stored in a system is kept constant, then as the discharge time decreases (or is pulsed), the delivered instantaneous power must increase proportionately. This results in massive instantaneous amounts of power being produced by pulsed power systems. In many cases, the peak power pulses of the system greatly exceed the average power [2]. When this happens, system components are at a high risk of damage from breakdown voltages and overcurrents. Special design considerations must be taken to mitigate these risks.

1.1.2. Applications of Pulsed Power

This fast delivery of energy (usually in the form of high-voltage or high-current) is useful in a variety of fields and applications. Medically, it has been used to treat cancer, power flash-x-rays, and provide gene therapy [1], [3], [4]. Pulsed power can be utilized in environmental applications by commercially producing ozone, preventing biofouling, and treating hazardous waste [1], [4]. It is harnessed industrially in material studies, by affecting the electro-plasticity of materials, powering plasma-based ion implantation, and annealing surfaces [1], [3], [5]. While pulsed power has a widely useful application space, it is especially pertinent in national defense applications of Directed Energy (DE). While the United States has been involved in DE research for over 50 years, the Department of Defense has made a concerted effort to field DE capabilities in the last decade [6]. Among these DE systems are high-energy lasers (HELs), high powered microwave (HPM) devices, and particle beams [1], [6]. The push to translate abstract research efforts into concrete developed systems has opened a market that this book' design seeks to attach to. Radiance Technologies, a private engineering contractor, has multiple teams specializing in these DE fields. The Microwave Technologies and System Development Operation (MTSDO) at Radiance Technologies builds prototype HPM pulsed-power systems for the U.S. government. A HPM system is often broken down in four subsystems, depicted in Figure 1.1 [7].

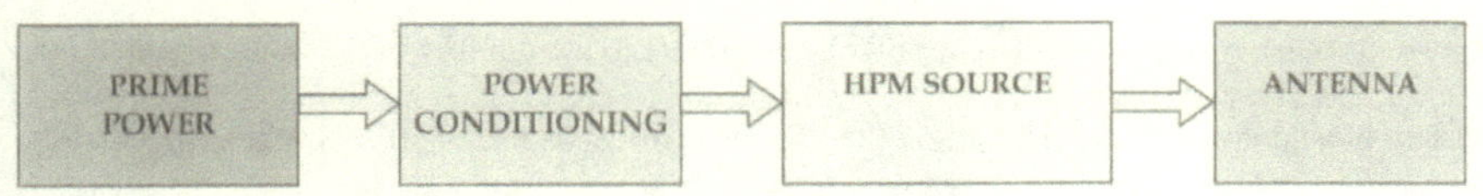

Figure 1.1 HPM system architecture [7]

First, the Prime Power subsystem comprises the storage of energy to be radiated by the HPM Source and Antenna. The Power Conditioning subsystem takes the stored energy and makes it usable to the HPM Source, both by voltage conversion and pulse shaping. The HPM Source takes the transformed energy and creates the high-power microwave by either drastically oscillating or amplifying the voltage. Finally, the Antenna radiates the high-power microwaves at the desired frequency and gain to produce an effect in the target. Of the 4 HPM subsystems, this book specifically outlines the design and implementation of a Power Conditioning HPM subsystem developed for MTSDO at Radiance Technologies.

1.1.3. Capacitor Charger Application

For the energy in a pulsed power system to be released quickly, it must first be contained in a type of energy storage system. Both capacitive and inductive energy storage systems can be used to store and deliver pulsed power, but capacitive energy storage is the more conventional and popular option. These capacitive energy storage systems can consist of a single capacitor, but often take the form of a capacitor bank, either connected in series or parallel [8]. Typically, the capacitor banks are charged up slowly until they hit a desired voltage, and then a switch is closed, and the energy is released instantaneously. However, there are certain applications where it is desirable to rapidly charge and discharge these capacitors in a repeated pulsed manner [9], [10]. In these situations, a special kind of capacitor charger topology is needed to meet these quick charge and discharge times. The repeated pulsed manner in these specific applications is known as pulse repetition frequency (PRF). PRF refers to the number of pulses transmitted by a system in a single second [11]. Setting the PRF is a critical aspect of HPM systems specifically, as it dictates the frequency that the HPM source can repeatedly fire and produce desired effects on the target. For the application of this book, making sure the capacitor charger can charge and discharge at a specific PRF is a critical design consideration.

1.2. Motivation

As mentioned in 1.1.2 and 1.1.3, the main motivation of this book comes from the desire to create a high-repetition rate pulsed-power capacitor charger for use in an HPM system. The system load will be a spiral generator used to create HPM. For the purpose of system evaluation, the specific type of spiral generator used can be simulated as a capacitor, since it stores energy in the same way as a capacitor [12]. Based on the desired packaging and use case of the device, an extremely compact and power-dense form factor is required. This consideration is foremost in the project, and thus special care must be taken in selecting the power topology, charger configuration, and switching devices used in the final design. Table 1.1 details the desired customer specifications to which the system needs to be designed for an effective HPM prototype.

Table 1.1 Required System Specifications

Design Parameter	**Required Value**
Input Voltage	400 - 500 V
Output Voltage	≥ 1200 V
Load Capacitance	0.47 μF
Charge Time	< 33 μs
Average Power	≥ 10.25 kW
Pulse Repetition Frequency (PRF)	≥ 30 kHz
Switching Frequency	1 MHz
# of Consecutive Capacitor Charges	300
Idle/Recharge Time	1-5 s
Size	4 lbs., 80 in^3

1.2.1. System Architecture

When it comes to DC-DC conversion, there are two main conventions engineers can choose: linear regulators or switch-mode power supplies (SMPS). Of the two options, linear regulators are much simpler to create and implement, but they come with major tradeoffs. Since they rely on controlling the voltage drop across their components, whether a simple resistor or a transistor, they can only output a voltage level lower than the input. And since the voltage drop across the components is generally thermally dissipated, linear regulators are very inefficient. While more complicated and expensive to create, switch-mode power supplies offer much more flexibility in voltage output, and are significantly more efficient [13]. In the application of this book, the overall system needs to be compact, efficient, and needs to produce a voltage output higher than the input. Thus, the SMPS is the clear choice for the general system architecture.

1.2.2. Power Topology

There are many different power electronics topologies under the SMPS umbrella. The design constraints of this book quickly narrow down the competition. The converter in question needs to be isolated to protect the primary side from EMI produced by the load on the secondary side. This constraint eliminates the use of non-isolated converters like the buck, boost, and Cuk topologies [14]. This leaves flyback, forward, half-bridge, full-bridge, and push-pull converters as viable options for this design. Among these isolated topologies, the half-bridge and full-bridge converters suit the application of this book, due to the low switch strains and high power levels available [15]. Of the two, the half-bridge converter has the lower output power. The full-bridge converter, while more complicated and expensive to design, can have higher output power than the half-bridge due to the switch currents being shared across two bridges. The circuitry for a standard full-bridge converter is shown in Figure 1.2.

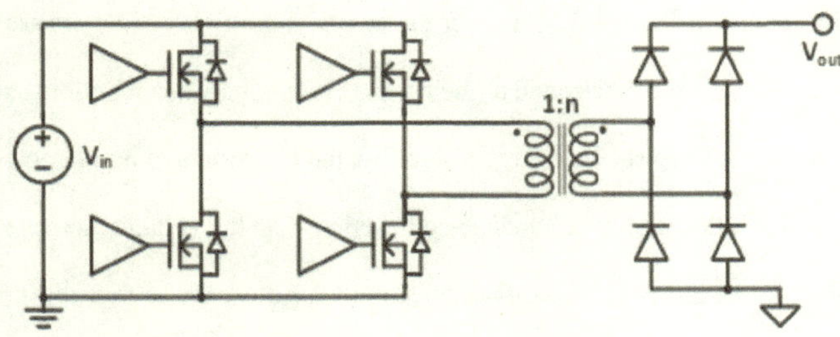

Figure 1.2 Full-bridge converter

1.2.3. Modularity

As shown in Table 1.1, the system to be designed must be able to take an input up to 500 V and output ≥ 1200 V. There a few different methods of stepping up voltage using full-bridge converters; the primary way is to use a step-up transformer. This process requires setting a greater than 1 transformer turns ratio, but this can cause switch voltage ringing and thermal strain, shown in [16]. To avoid adding stress to the switches in a converter, and to add to the system's stability, multiple converters can be aligned in four different configurations: Input-Parallel Output-Parallel (IPOP), Input-Series Output-Series (ISOS), Input-Parallel Output-Series (IPOS), and Input-Series Output-Parallel (ISOP) [17]. Each system configuration has its own merits, summarized in Table 1.2.

Table 1.2 Summary of Series and Parallel Converter Connection Schemes [18]

	Main application situation	**Key issues**
ISOP	Transfer high voltage to low voltage with power expansion	Voltage balance in input side, and current sharing in output side
ISOS	High input/output voltage	Voltage balance in output/input side
IPOS	Transfer low voltage to high voltage with power expansion	Voltage balance in output side
IPOP	Power expansion	Current sharing of output current

As shown above in Table 1.2, whenever the inputs of multiple converters are connected in parallel, power expansion happens due to the current sharing between individual systems. If the output current needs to be shared for extra stability, the converters' outputs can be connected in parallel again. But if the outputs are connected in series, then the voltages across each card are referenced to each other, allowing for voltage stacking. Figure 1.3 shows four full-bridge converters stacked in an IPOS configuration.

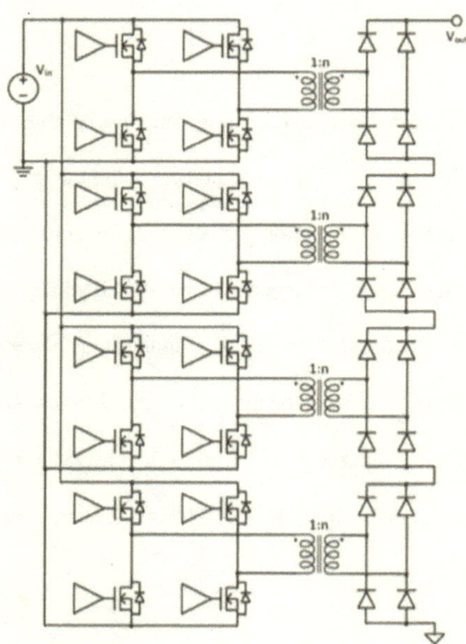

Figure 1.3 Full-bridge converters stacked in IPOS configuration

Since the system being designed in this book requires a higher output voltage than input voltage, the IPOS configuration is ideal. It allows the same 500 V input to be shared across the converters, resulting in lower switch stress than by using a single converter with a step-up transformer. The series output allows for voltage stacking scaled by the number of full-bridge converters in the system. Since a desired output voltage of ≥ 1200 V is required, using four 1:1 full bridge converters will allow the system to meet and exceed the voltage requirements with low stress on any individual switching bridges.

1.2.4. Semiconductor Material

In the last few decades, the introduction and implementation of wide bandgap (WBG) materials into semiconductor devices has drastically changed the achievable efficiencies and form factors of power electronic converters. Among these WBG devices, Silicon Carbide (SiC) and Gallium Nitride (GaN) have emerged as frontrunners [19]. When compared with traditional silicon (Si) switches, their desirable material properties become clear, as shown in Figure 1.4.

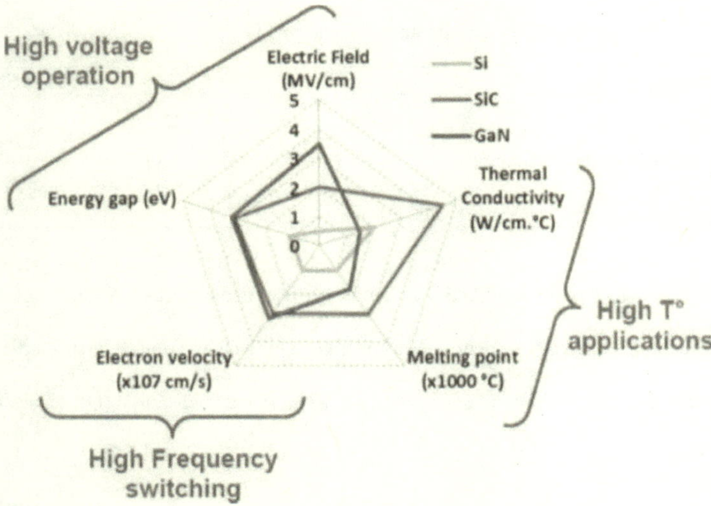

Figure 1.4 Comparison of different semiconductor materials [19]

Of the material properties shown in Figure 1.4, high voltage operation and high switching frequency are the most important to this research. Since pulsed power converters require high currents and voltages, the switches used must be able to hold off a high drain-to-source voltage (V_{ds}). Also, the extremely fast microsecond charge times require even faster rise times of the switching gate signals, at the nanosecond scale, requiring high switching frequencies in the Megahertz range. For both reasons, WBG semiconductor devices are ideal for use in pulsed power applications compared to traditional Si devices.

Between the two available WBG switches, GaN and SiC, GaN is particularly suited for compact pulsed power applications. While SiC switches have a higher thermal conductivity, GaN is suited for running at very high voltages and frequencies due to their high electron mobility and critical field, as shown in Table 1.3 [20].

Table 1.3 Material Properties of Silicon, GaN, and SiC [20]

Parameter		**Silicon**	**GaN**	**SiC**
Band Gap E_g	eV	1.12	3.39	3.26
Critical Field E_{Crit}	MV/cm	0.23	3.3	2.2
Electron Mobility μ_n	$cm^2/V \cdot s$	1400	1500	950
Permittivity ε_r		11.8	9	9.7
Thermal Conductivity λ	$W/cm \cdot K$	1.5	1.3	3.8

When it comes to driving these WBG switches, GaN semiconductors have the edge as well. Figure 1.5 illustrates that GaN switches have lower switching charge requirements and a faster switching transition due to their lower on resistance (R_{DSon}) and gate charge (Q_g) values [21].

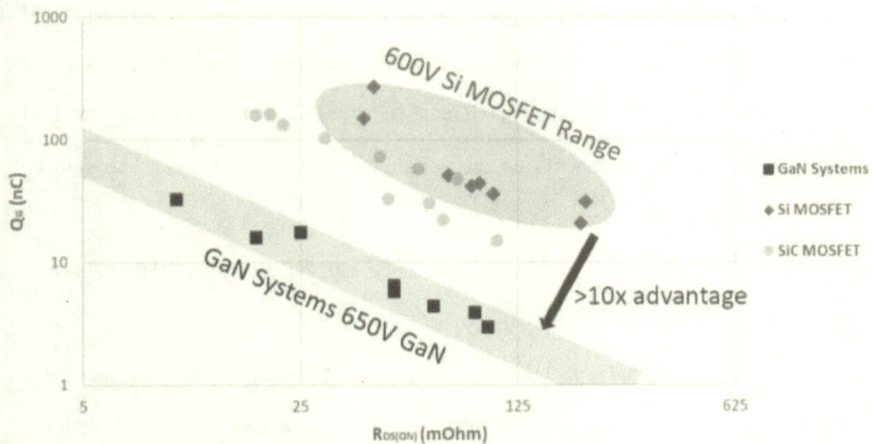

Figure 1.5 GaN switches vs. the competition [21]

While GaN semiconductor materials are a relative newcomer to power electronics switching applications, their beneficial characteristics over the competition make them a worthwhile pursuit for the purposes of this design. In the last decade, many implementations of GaN-powered converters have been successfully utilized and vetted [22], [23], further solidifying the case for using this semiconductor material in this project.

1.3. book Organization

The subsequent chapters of this book are laid out as follows:

CHAPTER II is a literature review, seeking to study and review similar designs to the one discussed in this book. It will look specifically at other capacitor charging systems, with an emphasis on pulsed power applications. In particular, it will study other IPOS implementations of pulsed power capacitor chargers, seeking to show how this design is unique compared to currently existing systems.

CHAPTER III discusses the design of the full-bridge converter, diving into the relevant mathematics involved with converter design, and outlining the specific parts selected to meet the system requirements. It will also contain preliminary circuit simulations of the specific parts to understand how they work together in the total design, followed by a study of the energy losses specific to a pulsed power system.

CHAPTER IV will show the finished PCB along with the test setup used to evaluate it. The test results of the finished system will also be discussed in this chapter.

CHAPTER V concludes the book with an evaluation of how the system performed as set by the design constraints and discusses the future work that can be done based on this design, with any improvements that could be made in the future.

CHAPTER II

LITERATURE REVIEW

This literature review seeks to provide a survey of the two main research areas applied in this book: IPOS applications of power electronics converters, and pulsed power capacitor chargers. A case will be made for the novelty and uniqueness of the application of this book.

2.1 IPOS Converters

As discussed in 1.2.3, the use of multiple power converters in series or parallel configurations can provide huge benefits to a power system. The study in [17] summarizes these benefits in a comparison of the four possible architectures (ISOS, ISOP, IPOP, and IPOS). Since each module in these configurations shares the total system power, the system has reduced electrical and thermal stresses. Also, the output power can easily be expanded by adding more modules to the system. Of the four configurations discussed in [17], the IPOS architecture becomes the clear candidate for this book, due to its predominant usefulness in high-voltage applications. Many different implementations of converters in IPOS configurations have been developed and studied using different types of converter topologies as their base module. The researchers in [24], [25] show the use of two individual push-pull converters in IPOS configuration for photovoltaic (PV) applications. Gruner *et al.* in [26] uses IPOS-connected forward converters to step up 30 V to 400 V. While the list of different kinds of converters connected in IPOS is extensive, this review will focus specifically on IPOS-connected full-bridge converters, since that is the chosen power topology for this book. An example implementation of full-bridge converters connected in an IPOS configuration with one common filter is shown in Figure 2.1.

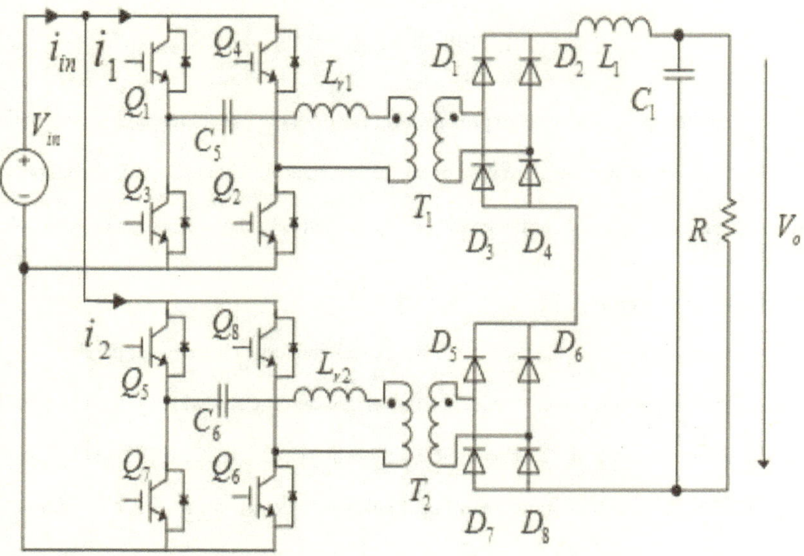

Figure 2.1 Full-bridge converters connected in IPOS with common filter [27]

As discussed in 1.2.2, full-bridge converters can produce higher power outputs than other converters due to their switching loads being shared across two branches. Thus, combining full-bridge converters in IPOS configuration, like in Figure 2.1, theoretically allows for very high sustained power levels. This becomes useful in a host of charging applications, where a high power level over a certain amount of time allows for faster charging. While different in base converter topology than Figure 2.1, Zhang *et al.* in [28], [29] show this point in the specific application of onboard EV chargers. Both take an input voltage of 24 V, and produce output powers of 1 kW [29] and 3.3 kW [28]. They both illustrate the voltage gains and high-power levels achievable by IPOS full-bridge converters. However, they differ drastically in power level, switching frequency, and application from the system this book seeks to design. As was discussed in Table 1.1, the proposed system requires a switching frequency of 1 MHz, and an average power of greater than 10 kW. Furthermore, the application of this book focuses on rapidly modulated pulsed charging of a capacitor, where most existing IPOS applications deal with continuous charging; whether an electric vehicle [28], [29], [30], photovoltaic array [24], [25], [31], or wind turbine [32], [33], [34]. Although these applications, and many more, utilize the multiple benefits of the IPOS configuration [17], there are very few existing designs that utilize an IPOS configuration in the specific context of pulsed power. One such design is discussed in [35], where the researchers seek to improve upon a high-voltage microsecond pulsed power supply (HV-MPPS), by implementing an IPOx configuration. IPOx means that the inputs are connected in parallel, but the outputs can either be connected in series or in parallel, depending on the desired power output. The specific system in [35] is used to power a plasma dielectric barrier discharge (DBD) reactor. While this type of pulsed application is more applicable to the one proposed in this book, a closer look reveals some key differences. First, the system architecture and power topology of a single module is quite different from a 1:1 full bridge converter. The HV-MPPS module uses a modified flyback converter with a high-gain 1:30 transformer turns ratio, pictured schematically in Figure 2.2. The high gain transformer on the flyback converter module allows for impressive voltage scaling; each module takes a 30 V input and outputs 7 kV. Combined in IPOS configuration, the system far exceeds the output voltage requirement of this book. However, the PRF of the flyback converter in the HV-MPPS is only 1.3 kHz, and the system charges up in 83 μs, values far beneath the 30 kHz pulse repetition frequency and 33 μs charge time set in Table 1.1.

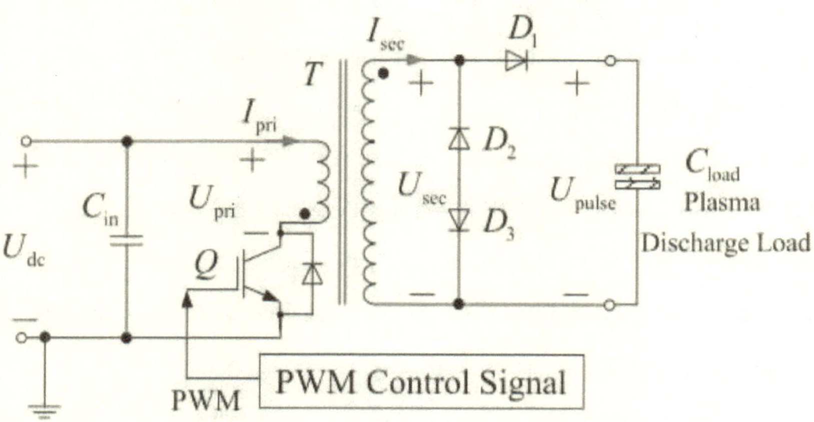

Figure 2.2 HV-MPPS flyback module designed in [35]

The researchers in [35] certainly show the value of using an IPOS converter configuration in the field of pulsed power, as opposed to the continuous power applications discussed above. But their design varies from this book in many critical areas, especially when it comes to application requirements. Thus, a further look at the specific application space of this book, namely pulsed power capacitor chargers, will be discussed in the following section.

2.2 Pulsed Power Capacitor Chargers

The basics of pulsed power capacitor chargers were covered in 1.1.3. To reiterate, pulsed power relies on the rapid discharge of stored energy. Capacitors are a form of energy storage that are particularly suited to this space, due to their ability to be charged up quickly, and rapidly discharge when shorted. There are applications where the capacitor(s) need to be charged up slowly and only discharged once, like the 3000 kV impulse generator in the High Voltage Laboratory at Mississippi State University [36]. But these applications are mainly suited for testing environments and come with their own design challenges. Since they greatly differ from the application space of this book, they will not be discussed in this review.

As discussed in 1.1.3, pulsed power capacitor chargers usually refer to systems required to rapidly charge and discharge at a certain pulse-repetition-frequency (PRF); these types of rapidly pulsed capacitor chargers are closer to this book application and will be reviewed. A closer look reveals the fact that while multiple existing pulsed capacitor chargers exist, few fit the specific system requirements outlined in Table 1.1. An example of this is in [9], where the researchers created a system capable of charging a 0.25 μF capacitor up to 20 kV in 3 ms. Compared to the single-shot system in [36], the results in [9] are more translatable to this book requirements. The system architecture, shown in Figure 2.3, consists of an input rectifier, inverter, transformer, resonant inductors and capacitors, and a high voltage output rectifier. The input rectifier is composed of three large thyristor modules, and the inverter consists of three IGBT modules. Many pulsed power systems use these large chassis mount switching modules instead of individual switches. Using these modules allows for an easy design process and quick integration, but restricts the amount of customization, and greatly increases the volume of the system.

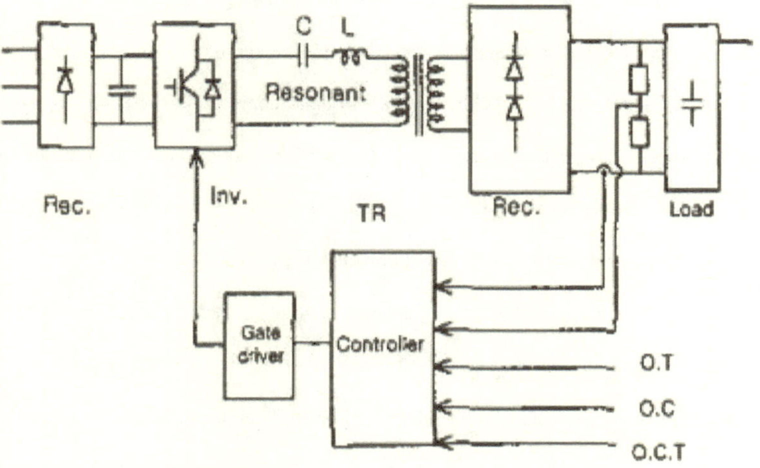

Figure 2.3 System configuration block diagram discussed in [9]

Holt *et al.* illustrates this point further in [37]. The researchers use a commercial-off-the-shelf (COTS) IGBT H-Bridge module in combination with a 1:170 step up transformer. Additionally, the output rectifier uses five diode bridge modules. While these sorts of systems can produce large amount of pulsed power (the system charges up to 50 kV in 1.5 ms in [37]), they do not fit the needs of this book. The use of COTS switching modules work well in benchtop designs, but they greatly exceed the form factor requirements in Table 1.1, as these individual modules are large and heavy. Figure 2.4 shows the packaging of the IGBT switching module used in [37]; an Infineon FF600R06ME3 [38]. One of these half-bridge modules is equivalent to ~10 in^3, and two are required to make a full-bridge converter. Thus, using modules like these for the design in this book would take up the entire form factor budget (80 in^3) with switches alone.

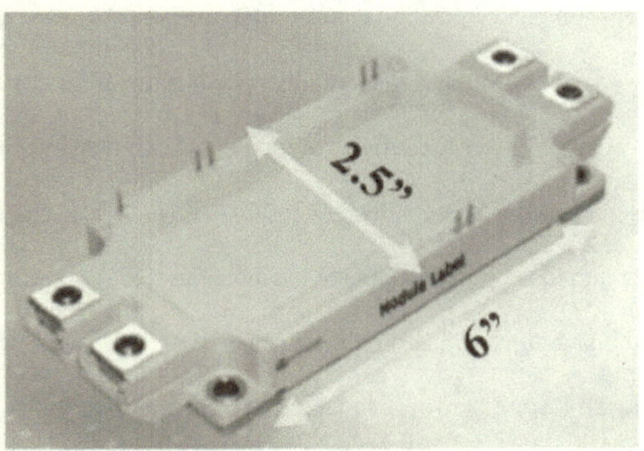

Figure 2.4 An example of an IGBT switching module [38]

Beyond the large size of the switches in [9], [37], the switching frequency of both systems is in the 10-20 kHz range, far beneath the desired 1 MHz switching frequency. This is another shortcoming of the larger IGBT switching modules; they have large power ratings but aren't designed to switch as fast as is necessary for this application. Zhang *et al.*, [39], describes a "compact" capacitor-charging power supply using a full-bridge series resonant charging circuit. However, a closer look shows that the system volume is over 3500 in^3 and the charge time is 10 seconds. While the final output voltage of the capacitor is a high 10 kV, the size and charge time fit a very different capacitor charging application than this book. The design in [40] is even larger volumetrically than [39], but it does repeatedly charge at a PRF of 1 kHz. This faster repeatable charge time is a step in the right direction but is nowhere close to 30 kHz. A similar design is shown in [41], where the system takes a 513 V_{dc} input and uses a full bridge converter to produce a 10 kW output power at a 42 kHz switching frequency. It is very large (~1240 in^3), and cannot repeatedly charge at a fast PRF, making it unsuitable for this book.

Finally, a review on pulsed power capacitor chargers would be incomplete without acknowledging the work of Dr. Michael Giesselmann. He is a major faculty member at the Center for Pulsed Power and Power Electronics at Texas Tech University, where his team has created multiple pulsed capacitor charger designs [10], [42], [43], [44]. The listed systems are designed to suit different pulsed power applications, including HPM generators like this book. They all charge up to high output voltages on the scale of 10 – 50 kV in the time scale of 5 – 50 ms. Additionally, they all feature hardware in large benchtop configurations, and utilizing large H-bridge IGBT modules like the one in Figure 2.4. Thus, when compared to the requirements of this book listed in Table 1.1, these designs run into the same issues as the systems earlier in this section; they are too large and too slow.

As has been shown, most of the existing pulsed power capacitor chargers offer a general function similar to the requirements of this book: a modulated, high voltage, "fast" charging of a capacitor. But these systems simply are not small or fast enough to be suitable. Many of these designs incorporate novel and useful concepts to suit their specific tasks; a variable output voltage in [41], improved burst mode efficiency in [37], etc. Some of them even make use of a full-bridge topology. But none of them use full-bridge converters connected in an IPOS configuration, utilizing individual GaN semiconductors for switching.

2.3 IPOS Pulsed Power Capacitor Chargers

So far, this literature review explored existing power electronics systems that incorporated a modular IPOS configuration, specifically using full-bridge converter topologies. Connecting converters in IPOS has clearly benefitted many systems, but none of them are close in application to a pulsed power capacitor charger. Next, the review looked specifically at the different pulsed power capacitor chargers that currently exist. While closer in application to this book, these existing systems generally were either too large or too slow to fit the specific requirements laid out in Table 1.1. On top of that, none of them utilize the specific full-bridge IPOS configuration that this book seeks to explore. In the end, only one existing system was found to implement the specific power topology of this book' proposed system in a pulsed power capacitor charger application. This system will be discussed in the following paragraphs.

The paper in which the system is discussed is [45], *Analysis and Design Considerations of Input Parallel Output Series-Phase Shifted Full Bridge Converter for a High-Voltage Capacitor Charging Power Supply.* This design checks the two major boxes of this review: an IPOS configured converter, and a high-voltage capacitor charger. The system architecture is shown in Figure 2.5.

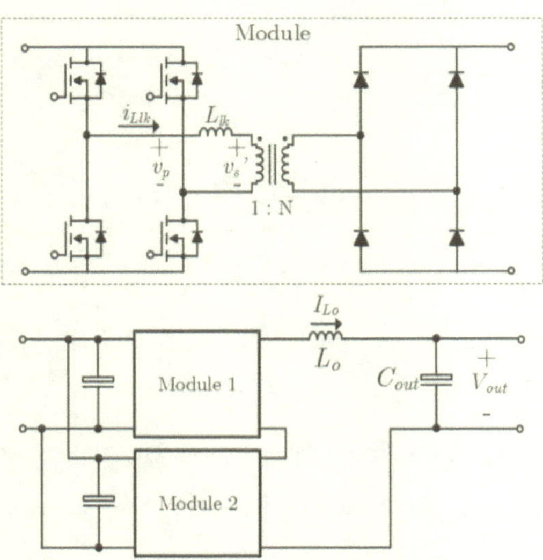

Figure 2.5 IPOS topology used in [45]

As shown above, the system in [45] looks very similar to the proposed architecture and topology discussed in 1.2. An individual full-bridge converter is modularly connected in IPOS with another converter to charge a capacitor C_{out}. A single inductor L_o is used to filter the system output current. One aspect specific to this design is the use of phase-shifted full bridge converters, opposed to classic full-bridge converters. Phase shifted full bridge converters implement zero volt switching (ZVS) which allows for lower switching losses [46]. Not only does the system in [45] have a similar architecture to this book, a similar charging cycle is outlined for this system, shown in Figure 2.6.

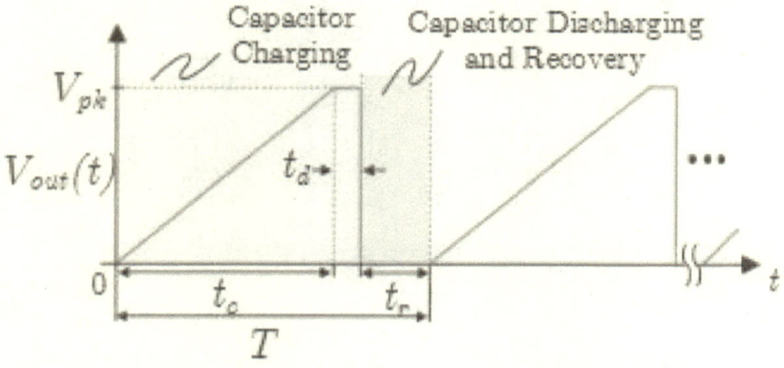

Figure 2.6 Output voltage cycle proposed in [45]

This output voltage cycle pictured above is evaluated in terms very comparable to this book. The system charges the capacitor in the charge time t_c, hits a desired peak voltage V_{pk} and stays there for a dwell time t_d, discharges, and rests for the recovery time t_r before starting the cycle over again. These 3 separate time periods are encapsulated in total charge period T. Taking the inverse of T gives the PRF of the system. Clearly, the proposed architecture and charging cycle of the system in [45] are directly translatable to the application of this book. But looking at the specifications for the final hardware setup shows where the two designs start to differ. Table 2.1 shows the final specs for the system in [45].

Table 2.1 Hardware Specifications in [45]

Parameter (Unit)	Value
DC-Link Voltage V_{dc} (V)	200
DC-link Capacitance C_{dc} (mF)	0.92
Switching frequency (kHz)	100
Output Inductance L_o (mH)	0.8
Output Capacitance C_o (uF)	5
Voltage peak V_{pk} (V)	1150
Pulse frequency f_p (Hz)	100
Charging time t_c (ms)	6
Dwell time t_d (ms)	1
Recovery time t_r (ms)	3

When compared to Table 1.1, Table 2.1 shows the key differences between this book and the system designed in [45]. The peak output voltage V_{pk} comes close to the $\geq$ 1200 V output voltage required, but it takes a much longer time to charge to this voltage than desired. The charging time t_c is 6 ms, two orders of magnitude slower than the < 33 μs charge time needed in this book. Moreover, the PRF (f_p) is 100 Hz, and so the system cannot repeatedly charge the output capacitor at the 30 kHz PRF. While there are other differences, like the output capacitance and switching frequency, the charging time and pulse frequency are too far off the mark to suit the requirements of this book. And lastly, the final hardware is too physically large to fit the required form factor. While not concretely specified in the paper, the hardware is pictured

(Figure 2.7) and in its current spread bench form looks much larger than the compact 80 in^3 set in Table 1.1.

Figure 2.7 Final hardware setup in [45]

Thus, despite having a similar system architecture and topology (IPOS full-bridge pulsed capacitor charger), the design discussed in [45] differs too greatly in final specifications to be suitable to the requirements of this book.

2.4 Summary

This chapter has sought to show the novelty of the design proposed in this book by reviewing the existing literature regarding IPOS converters and pulsed power capacitor chargers. While systems using an IPOS configuration have been well published, they have almost never been designed for a pulsed power capacitor charging application. And while multiple pulsed power capacitor chargers exist, none satisfy the specific design requirements laid out in Table 1.1, or use the same system architecture. Therefore, the system design will be outlined in CHAPTER III since it has been shown that the proposed design offers a unique solution to a unique set of requirements.

CHAPTER III

DESIGN & SIMULATION

3.1 Overview

As stated in CHAPTER I, the purpose of this book is to create a compact DC-DC pulsed power converter that can be connected in an input parallel, output series (IPOS) configuration. This configuration allows for modularity and scalability across a range of pulsed power applications, although this design is specifically tailored to charge a pulsed-power capacitive load used in a high-powered microwave (HPM) prototype. A maximum input voltage of 500 V is required for the module, with a 1:1 transformer voltage conversion ratio producing a 500 V output. The output voltage can be scaled up by combining modules in the IPOS configuration; this design focuses on combining a maximum of four converter modules. A standard full-bridge converter will be used as the base module, with a LC filter on the input. The following sections detail the mathematical analysis, part selection, and simulation processes used to inform the final design.

3.2 Mathematical Analysis

Before selecting the specific parts to be used in the converter, a mathematical study needs to be conducted to analytically determine the load on the converter and its respective parts. Since the converter is driving a purely capacitive load, managing inrush currents that could cause damage to individual components is a key consideration. An extra element must be added in series with the load capacitor to mitigate this risk, as discussed in [47]. Adding a properly sized inductor will help filter out a majority of the inrush current. Another way to minimize inrush current on top of filtering is by careful control of the switching duty cycle. Known as a "soft start" process, the duty cycle is gradually increased throughout the entire charge, which in turn gradually increases the output power of the system [47]. Figure 3.1 shows the filter inductor and soft-start operation starting at the converter output.

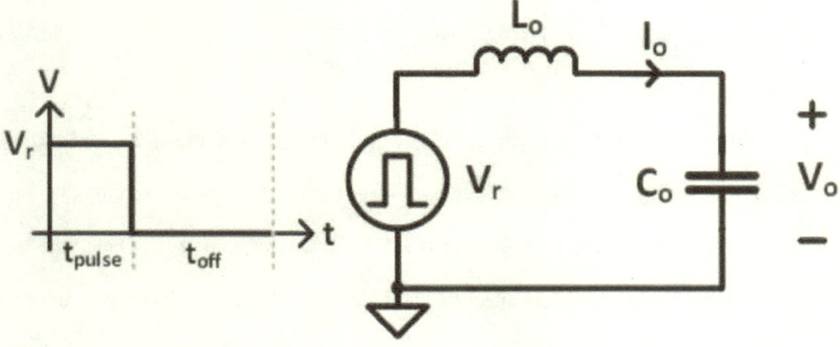

Figure 3.1 Model of output side of converter [47]

Since the inductor and capacitor in series on the converter output form a resonant circuit, resonant circuit equations can be used to determine load on the system and thus proper component sizing. Equations 3.1 – 3.4 describe the resonant behavior seen by the converter output. The variable *i* in the equations iterates through each switching cycle within the 33 µs charge window set by Table 1.1. Furthermore, the C_O value was set to be 0.47 µF according to the system requirements in Table 1.1. Based on this value, the L_O inductance must be in scale of hundreds of nH to keep the cutoff frequency of the LC filter less than the 1 MHz switching frequency of the system. The value was set to 0.47 µH, a common inductance value in this range.

$$V_{o\_on}[i] = \sqrt{\frac{L_o}{C_o}}\, I_{o\_off}[i-1]\, sin\left(\frac{t_{pulse}[i]}{\sqrt{L_oC_o}}\right) + (V_{o\_off}[i-1] - V_r)cos\left(\frac{t_{pulse}[i]}{\sqrt{L_oC_o}}\right) + V_r \tag{3.1}$$

$$I_{o\_on}[i] = \sqrt{\frac{C_o}{L_o}}\, (V_r - V_{o\_off}[i-1])\, sin\left(\frac{t_{pulse}[i]}{\sqrt{L_oC_o}}\right) + I_{o\_off}[i-1]cos\left(\frac{t_{pulse}[i]}{\sqrt{L_oC_o}}\right) \tag{3.2}$$

$$V_{o\_off}[i] = \sqrt{\frac{L_o}{C_o}}\, I_{o\_on}[i]\, sin\left(\frac{t_{off}[i]}{\sqrt{L_oC_o}}\right) + V_{o\_on}[i]cos\left(\frac{t_{off}[i]}{\sqrt{L_oC_o}}\right) \tag{3.3}$$

$$I_{o\_off}[i] = \sqrt{\frac{C_o}{L_o}}\, (-V_{o\_on}[i])\, sin\left(\frac{t_{off}[i]}{\sqrt{L_oC_o}}\right) + I_{o\_on}[i]cos\left(\frac{t_{pulse}[i]}{\sqrt{L_oC_o}}\right) \tag{3.4}$$

In reference to Figure 3.1, Equations 3.1 and 3.2 show the output voltage and current when the rectifier voltage is V_r for the duration of t_{pulse}. Equations 3.3 and 3.4 show the output voltage and current when the rectifier voltage is at 0 V for the duration of t_{off}. Due to the "soft start" switching profile to be implemented, the value of t_{pulse} increases with every iteration of *i*, as the switching period is gradually increased. When Equations 3.1 – 3.4 are solved under the constraints given in Table 1.1, voltages and currents on the secondary side of the transformer are given. The maximum secondary voltage is 405.319 V and the maximum secondary current is 127.881 A. Since the design uses a 1:1 transformer, the same maximum voltage and current applies on the primary side. These values found from the output resonant equations give us a voltage and current threshold that can inform the part selections in the proceeding sections.

3.3 Part Selections

The following sections outline the selection of the major parts of the converter, based on the requirements set in Table 1.1, the analysis conducted in 3.2, and general electrical engineering principles.

3.3.1 FET's

As discussed in 1.2.2, there are two pairs of switches in a full-bridge converter (four total) that alternate opening and closing. As mentioned in 1.2.4, the decision to use GaN-based semiconductors for these switches was made after an evaluation of the currently available semiconductor materials on the market, due to their high switching frequency and voltage ratings. When it came to selecting the actual GaN switches, there were a few options to consider. Doing a survey of the current market for these switches, the main type of GaN-based switch widely in production is the GaN HEMT (high electron mobility transistor), which are designed to be a direct replacement for traditional power MOSFETs [20]. While a few different companies have produced GaN semiconductors in the last decade, GaN systems is a world leader specifically in the design, development, and production of GaN power semiconductors. Their portfolio of GaN semiconductors is split into two voltage ratings, 100 V or 650 V. Since the system being designed in this book has a required input voltage of 400-500 V, the V_{ds} of the switches must be able to withstand that voltage. Thus, only the 650 V rating is a viable product family for this application. Along with the V_{ds} rating, the high output power requirement demands a high I_{ds} rating. The GaN systems GS66516 transistor has the highest V_{ds} (650 V) and I_{ds} (60 A) rating of any compact switch in their lineup. Shown in Figure 3.2, there are two cooling options for this component, top-side cooled, or bottom-side cooled. Both are extremely compact chips, with a 11 x 9 mm^2 PCB footprint [48], [49].

Figure 3.2 GaN systems GS66516B [48] vs. GS66516T [49]

Of the two package options, the top-side cooling is the more desirable one, for its flexibility on cooling options down the line. During the testing and evaluation phase of the design, an external heatsink can easily be added to this part if it is found to be reaching thermal overload. Thus, the GaN systems GS66516T transistors are the switches selected for the full-bridge converter.

3.3.2 Gate Driver

Because of the high (1 MHz) switching frequency and voltages of the GaN HEMTs used in this project, special care had to be taken in selecting proper switch driving circuitry. A study of gate drivers suited for GaN HEMTs done by researchers in [50] outlines three major types of drive schemes currently available: conventional drivers, smart drivers, and resonant drivers. Of the three schemes discussed in [50], the smart driver is the ideal gate drive topology, offering galvanic isolation and protection between the input and output. Conventional gate drivers do not offer galvanic isolation and thus restrict high voltage switching applications. Resonant gate drivers are used in resonant converters and have yet to see major commercial production. Typically, smart gate drivers are designed to drive a half-bridge of two switches, and so two gate drivers are needed to drive the full bridge in the final design. An additional requirement of the gate driver was the ability to run two isolated channels to drive each gate, rather than a bootstrap configuration. This means that each output channel has its own supply voltage and are therefore isolated from each other. Figure 3.3 shows an example of a smart gate driver block diagram, with dual isolated output channels and galvanic isolation across input and output channels.

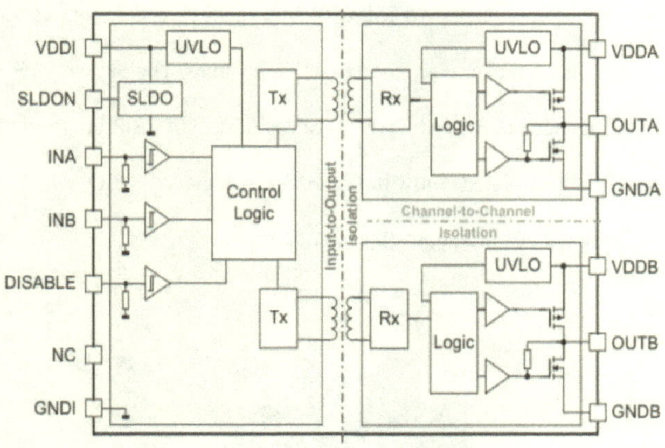

Figure 3.3 Smart gate driver block diagram [51]

Another requirement of the gate driver is having a low under voltage lockout (UVLO) value. Depicted above in Figure 3.3, the UVLO keeps the drive voltage from dropping below a certain value to avoid damaging the switches. Previous work done by Radiance Technologies in [47] found that having too high of a UVLO value on the gate driver can create brownout issues. Due to the system noise, and low bootstrap capacitance value, the supply voltage would drop low enough throughout the switching cycle to trigger the UVLO and stop the transistors from switching. A gate driver with a low UVLO value should mitigate this issue, but adding a dedicated power supply to each isolated voltage rail of the gate driver will ensure the drive voltages never drop too low. Lastly, as with all parts of this system, overall compactness is the main driving consideration. There are few options for gate drivers that meet all these requirements on the market. The two best options come from Skyworks and Infineon Technologies, the Si8274GB1-IM1 and 2EDF7275KXUMA1 parts respectively. Both come in very small DFN-style packages, support 1 MHz PWM switching, and feature galvanic isolation, dual isolated voltage rails on the outputs, and low UVLO values. The Skyworks Si8274GB1-IM1 has a lower UVLO value at 3 V, which is more desirable, but it suffers from stocking issues. The 2EDF7275KXUMA1, depicted below in Figure 3.4, has a 4 V UVLO, which should not be an issue with a 5 V isolated power supply on each output.

Figure 3.4 Infineon 2EDF7275KXUMA1 gate driver [51]

Consequently, two 2EDF7275KXUMA1 will be used in this design to drive the GS66516T half-bridges. Four isolated 5 V power supplies will be used on each gate driver output to sustain the drive voltages; the Würth Elektronik 1769205132 [52] has been selected for this task due to its compact form factor.

3.3.3 Input Filter

To ensure the power coming into the converter stays at a consistent and usable level, without any electromagnetic interference, special care must be taken in designing an input filter. When adding an input filter to a closed-loop system, Middlebrook's extra element theorem must be accounted for to keep the feedback loop stable [15]. The device being designed in this book, however, is an open-loop system. Since the system is rapidly charging a reactive load at a 30 kHz frequency, any feedback loop implemented would require a frequency in the 100s of kHz range. Design of such a fast and responsive loop would be very difficult and is unnecessary for this design. Thus, the only thing the input filter needs to account for in this design is providing a local impedance loop for the pulsed current being drawn through the switches. Without a filter, the supply voltage could experience sharp drops each switching cycle from the pulsed current and the system output power could drop. The filter will fundamentally be an LC filter, combined with an RL parallel damping circuit. The damping will make sure that the filter still attenuates the system's switching frequency, and not allow for instability at the resonant point. After simulating and evaluating different component values for the filter, the following values were selected, as shown in Figure 3.5. Figure 3.6 shows the waveform of the impedance profile of this input filter across the frequency spectrum. This filter design successfully dampens the resonant frequency to attenuate -15 dB at its peak. And the frequency that is most important, 30 kHz, it attenuated at -24 dB by this filter. For the inductors in the filter, a Vishay IHLP power inductor of the 240 nH value will be used, in parallel with a 24 nH Coilcraft inductor. The input filters' capacitance is split into four individual components: two 1 nF capacitors to be placed locally at each half-bridge switching branch, and two 2 μF high-frequency Ceralink capacitors.

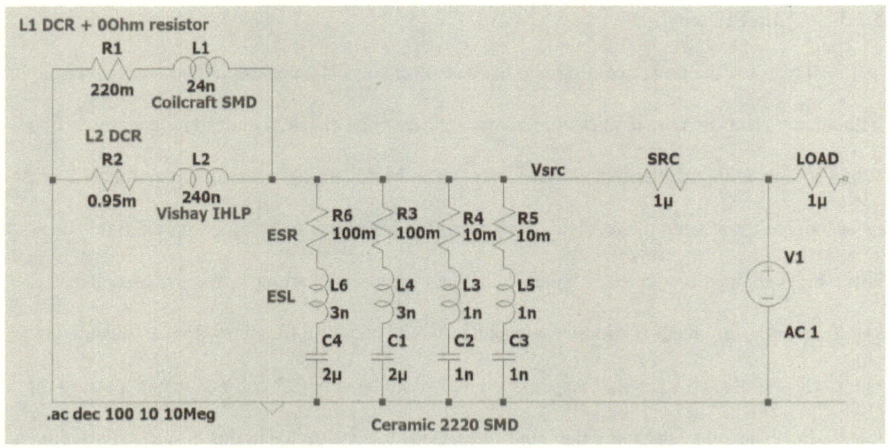

Figure 3.5 Input filter circuitry simulation

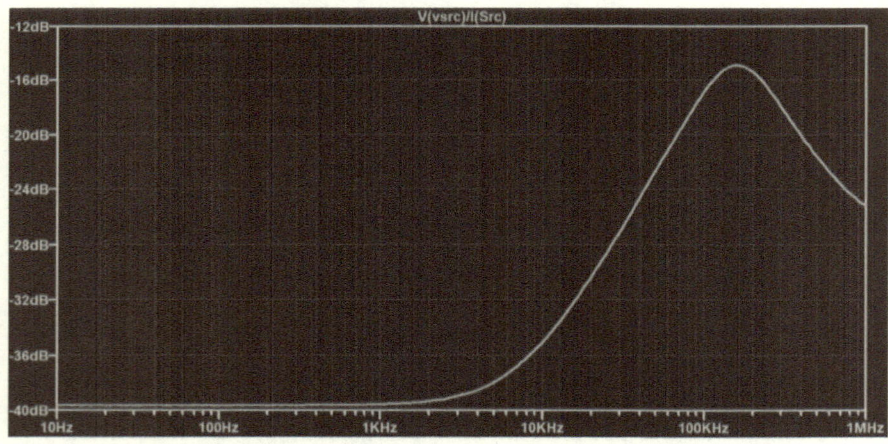

Figure 3.6 Input filter simulation impedance waveform

The Ceralink capacitors are ideal for this application due to their compact form factor and high-frequency optimization [53]. Most capacitors with a value this high are large through-hole or chassis-mount components that would far exceed the form factor requirements of this design. The Ceralink line from TDK use PZLT ceramic to make their capacitors for "fast-switching semiconductors [53]. Their compact surface mount form factor is pictured in Figure 3.7.

Figure 3.7 High-frequency Ceralink capacitor [53]

All the selected capacitors have a breakdown voltage > 500 V, the maximum input voltage of the system. To ensure the components work in the system as expected before production, they will be evaluated in the full system simulation in 3.4.

3.3.4 Transformer

With the specific pulsed power requirements of this book, it is too difficult to source a COTS transformer that fits the exact needs. Thus, a custom transformer will be created specifically to suit this design. Transformers are often the bulkiest part of any system, especially when custom made. But planar transformers are an appealing option, as they sit relatively flat, and thus work well in the stacked IPOS configuration of this book. Figure 3.8 shows the mechanical cross-section of a typical planar transformer.

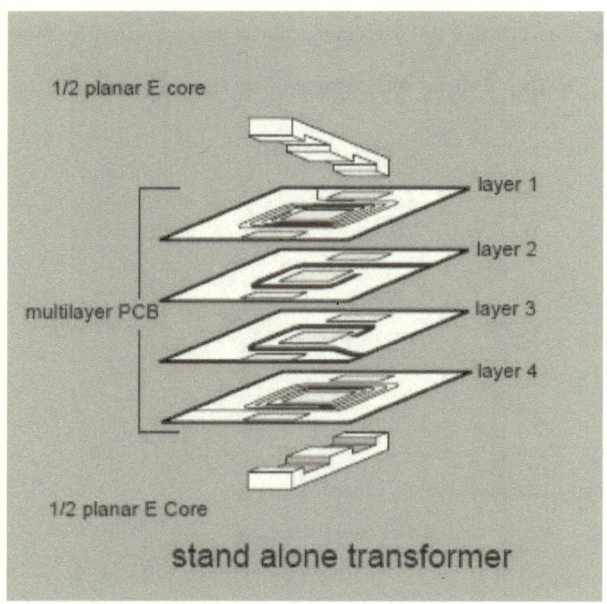

Figure 3.8 Example planar transformer [54]

As seen above, the primary feature of planar transformers is having their windings on a flat PCB, rather than wire wound around a bobbin. The core sandwiches the winding PCB, making the whole package very compact. When designing a planar transformer, the core shape, core size, and core material are major drivers in the design process. To decide which core to use in this design, a list was made of the commonly available planar core shapes and materials on the market. Each core has a list of values that can be used to determine the reluctance of the core, the reluctance of the air gap, and the primary and secondary inductance, when combined with the requirements given in Table 1.1. Based on these values, the number of turns needed on the transformer can be calculated. Another important calculation is ensuring that the transformer will not saturate during operation. This is done by calculating the B_{max} of the material, shown in Equation 3.5.

$$B_{max} = \frac{I_{pri\_max} \cdot N_p}{A_e \cdot (R_c + R_g)} \tag{3.5}$$

The B_{max} calculated in Equation 3.5 must stay beneath the B_{max} of the material, a value often found in the datasheet of a ferrite core material. With all these values being calculated for each transformer core, it became clear which option would be ideal for the design in this book. A balance had to be struck between the size of the core and the number of turns needed to meet electrical requirements. The smaller the core, the more turns were needed, while larger cores could pose issues to minimizing form factor. Since planar transformer windings are copper traces on a PCB, the number of windings that can fit within a core is much more restricted than a traditional wound transformer. The E32/6/20 core size in the 3F36 core material is a good balance of core size and winding requirements for this design [55]. It is 32 mm by 20 mm, with a 6 mm height, and only five turns are needed on the primary and secondary windings of the transformer. For the windings, a simple board with a single five-turn winding will be designed. This way, the same board can be used both for the primary and secondary side of the transformer. It only must be a two-layer PCB which is inexpensive and easy to manufacture. Figure 3.9 shows the PCB layout for the two-layer five-turn planar winding board.

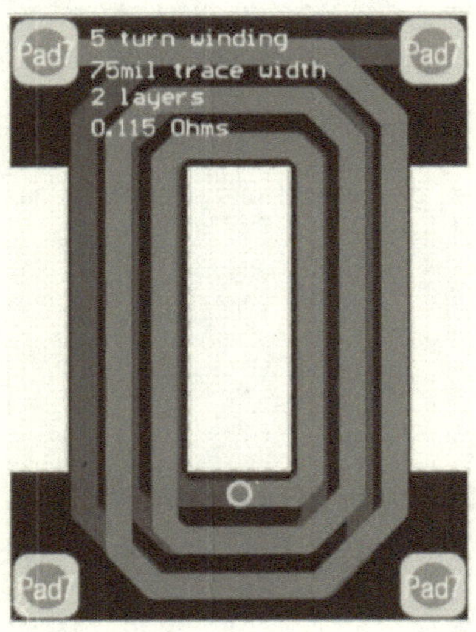

Figure 3.9 Planar winding PCB layout

The equivalent series resistance of the winding is denoted on the silkscreen of the board in Figure 3.9 to be 0.115 Ω. Keeping this value as low as possible is critical in mitigating a voltage loss across the transformer. The use of thick traces helps keep this ESR low. While some planar transformers are designed with their windings inside the main converter PCB, having a separate board allows for replacement of the transformer should it become necessary in the testing process, whether due to damage, or the desire to run at different electrical levels. The planar transformer is designed to be screw mounted instead of soldered on to the main PCB to further the ease the swapping process. When combined with the E32/6/20 ferrite core, the planar winding PCB boards are shown below in Figure 3.10.

Figure 3.10 Planar transformer 3D rendering

The custom design of this planar transformer will ensure that this design meets all its electrical requirements while remaining compact. It also allows for further customization of the converter, with the ability to easily change out the transformer with a higher or lower turns ratio to suit different applications.

3.3.5 Rectifier Diodes

Another critical part of any full-bridge converter is the rectifier that rectifies the AC on the secondary side of the transformer to DC. Four diodes are needed to construct a full-bridge rectifier, and there are a few special considerations to be made when selecting the diodes for this design. First and foremost, the diodes selected must be in a compact package. Secondly, they must be able to handle the high output power of the system. This means they must have a breakdown voltage greater than 500 V, and a surge current rating higher than 120 A, the values calculated in 3.2. Finally, they must have little to no reverse recovery time, due to the very fast switching cycles of the converter. GeneSic Semiconductor utilizes silicon carbide to develop semiconductor devices that can handle fast switching times at high voltages. Their MPS line of SiC Schottky diodes [56], [57], [58] are ideal for this design, as Schottky diodes are designed to have miniscule reverse recovery times. They have three package sizes in their offering that suit this system's size constraints: DO-214, TO-220-2, and TO-247-2, shown side by side in Figure 3.11.

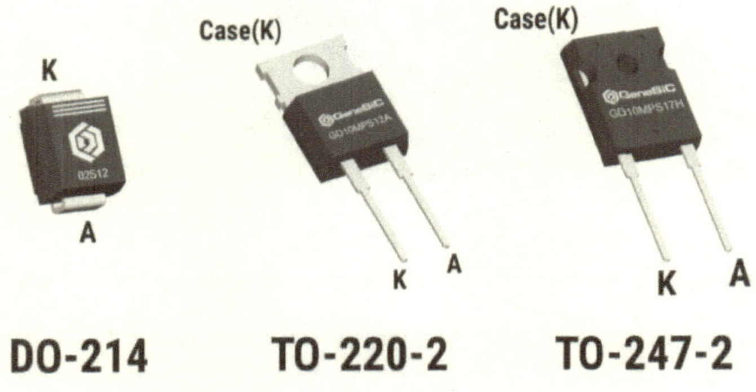

Figure 3.11 GeneSic SiC Schottky MPS diode packages [56], [57], [58]

Each option shown above meets the breakdown voltage requirements, with a minimum of 1200 V holdoff. The GB02SLT12-214 [56] in the DO-214 package would be ideal for this system due to its compactness, but it does not meet the desired current ratings, with a non-repetitive peak forward surge current ($I_{F,MAX}$) of only 100 A. The GD10MPS17H [58] in the TO-247-2 package has the highest power ratings of the three, with a 1700 V breakdown voltage and $I_{F,MAX}$ of 500 A. However, the TO-247-2 package is much larger and would easily be the largest component on the board, creating size issues. The GD10MPS12A [57] strikes a balance between the other two parts, with a smaller TO-220-2 package size than the TO-247-2, but greater power ratings than the small GB02SLT12-214 part. It offers a 1200 V repetitive peak reverse voltage and an $I_{F,MAX}$ of 400 A. Also, the exposed pad on the top and back of the package allow for easy application of a heatsink if it becomes necessary in testing.

3.3.6 Output Inductor

The final component to be selected on the converter before simulation and production is the output filter inductor. As discussed in 3.2, the converter is driving a purely capacitive load, which can create damaging inrush currents if left unmitigated. While using a "soft start" switching profile with help, a filter inductor must be added in series with the output capacitor. One large filter inductor on the output of the entire system might be beneficial in a large benchtop setup, like in [27], but it would be too large for the required form factor, and would have to be separate from the individual converter modules. Since this book' design centers around one single converter to be multiplied and combined in an IPOS configuration, it is necessary that each converter have its own filter inductor. The value of this inductor must be 0.47 μH, determined in 3.2, to ensure the filter's cutoff frequency stays beneath the switching frequency of the system. Many electronics companies produce power inductors, which have relatively high power ratings in compact surface-mount packages. Vishay Dale produces the IHLP line of power inductors, which come in a variety of package sizes, up to 67 mm^2 [59]. As the size of these inductors goes up, the power rating generally follow. The 50 mm^2 size is ideal for this application, since it has a smaller footprint than the 67 mm^2 line but keeps a saturation current above 60 amps. For the 0.47 μH part specifically, it has a 63 A DC saturation current. Thus, the IHLP5050FDERR47M01 is the selected part for the output inductor in this design, shown in Figure 3.12.

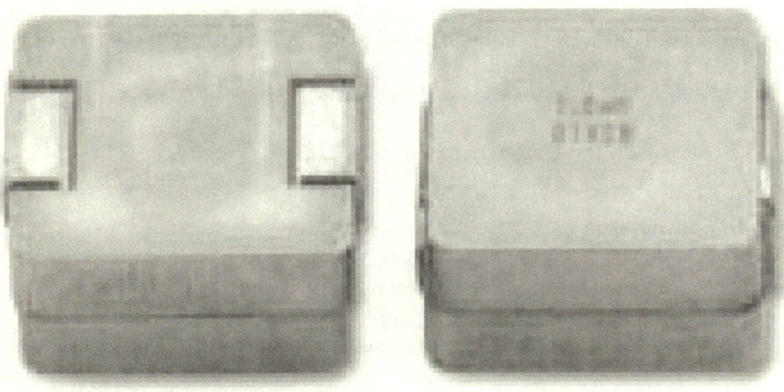

Figure 3.12 IHLP5050FDERR47M01 power inductor package [59]

Should the need arise in testing for a different value output inductor, whether the inductance value or power rating, replacement should be simple with the use of the common 5050 pad size.

3.4 System Simulation

To verify the design before production, an LTSpice simulation was created of the system. First, an individual full-bridge converter was simulated, utilizing components with characteristics as close to the actual selected hardware as possible. Figure 3.13 shows the simulated circuitry of a single converter in LTSpice. The goal of this initial simulation was to ensure that the selected system architecture and parts would succeed in charging up the load capacitor in the required 33 μs timeframe.

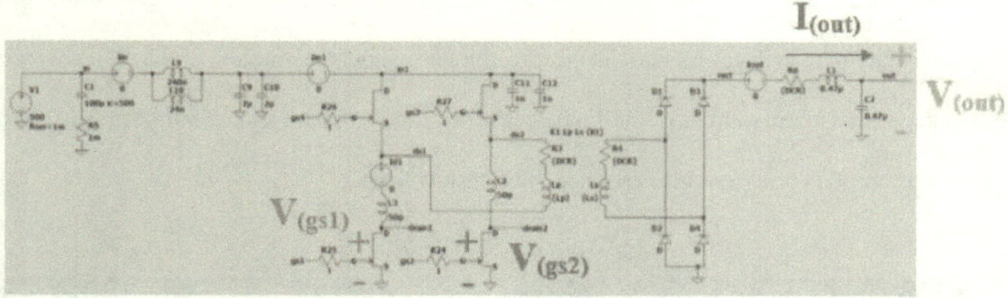

Figure 3.13 Single full-bridge converter simulation

Starting on the left of Figure 3.13, the simulation has a 500 V source voltage powering a 100 μF capacitor. The 100 μF was arbitrarily selected for the simulation to represent the input capacitor bank; it is a sufficient capacitance to charge the system once. The power then goes through the input filter, with values set as they were selected in 3.3.4. Since the switches are the most critical component in the system, the four switch models were sourced directly from GaN Systems' library, to ensure as accurate a simulation as possible. The transformer was set with a 1:1 turns ratio, and a DCR value of 115 mΩ was given to each turn. The output inductor was set to a value of 0.47 μH, reflecting the selection in 3.3.6., and the load capacitor was set to the system requirement of 0.47 μF shown in Table 1.1. To run the "soft start" switching profile, two piecewise linear files were created: one for the high side switching half-bridge, and one for the low side. These files dictate when the gate voltages should be driven high and low based on the 30 kHz PRF and the 1 MHz switching frequency. When the switching frequency is divided by the PRF and rounded down, a total of 33 individual switching cycles result. This means each switch period is 1 μs since the total charge time was set to 33 μs in Table 1.1. Over the 33 switching periods, the width is incremented each time until a full 40% duty cycle is reached on the final switching cycle. The result of this "soft start" profile is shown in Figure 3.14, as the two complimentary gate voltages increase in pulse width throughout the duration of the 33 μs charge. The voltage and current outputs of the single converter simulation are shown in Figure 3.15. See Figure 3.13 for the corresponding reference of each waveform.

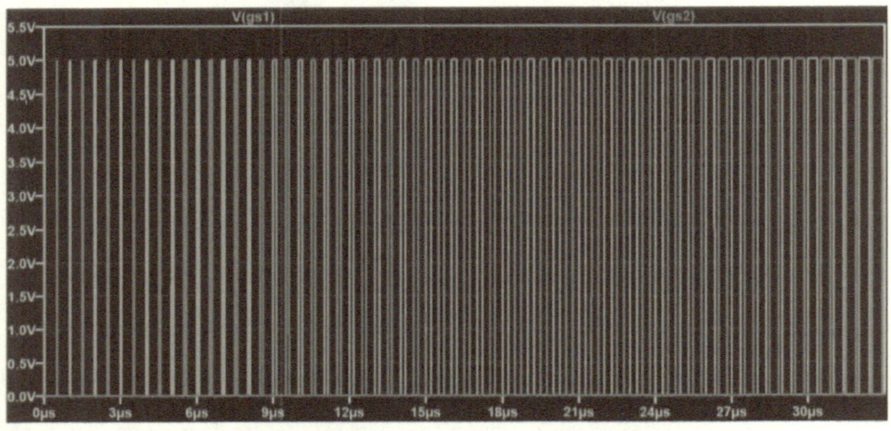

Figure 3.14 "Soft start" gate voltages

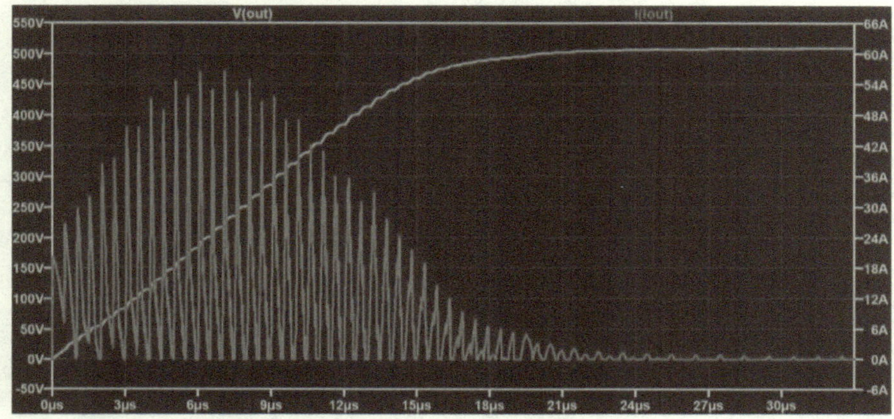

Figure 3.15 Single converter simulation output voltage and current

As shown in Figure 3.15, the simulation achieves the desired result; the output voltage fully charges up to 500 V in the allotted 33 µs timeframe. The output current also performs as expected, with peaks under 60 A. With the successful simulated run of a single converter, the full 4-converter IPOS simulation can be constructed. Figure 3.16 shows the circuitry for the full simulated system.

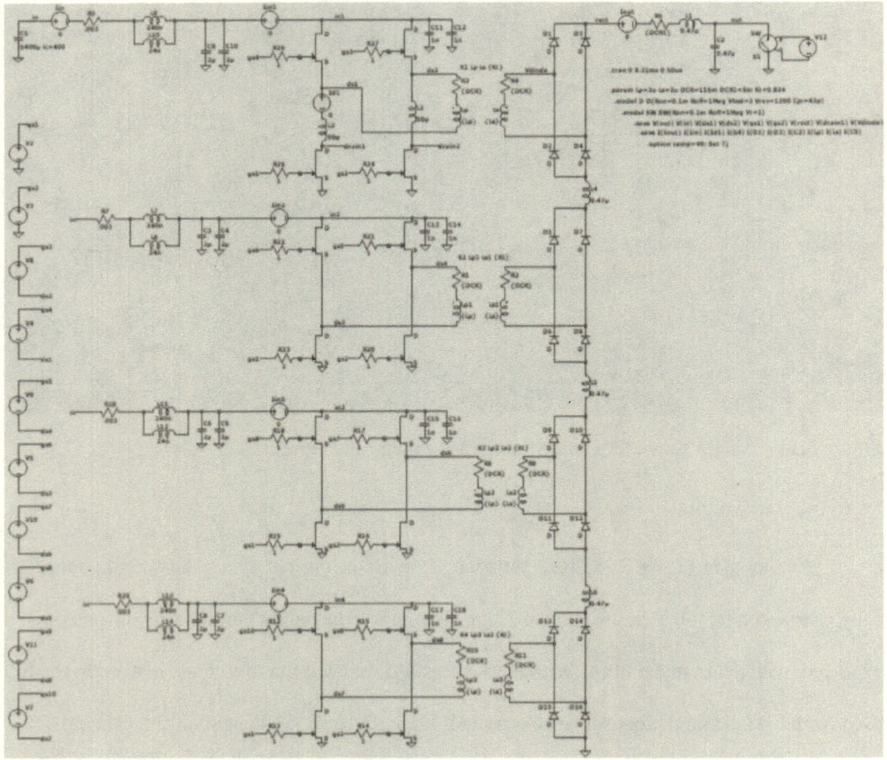

Figure 3.16 Full IPOS system simulation

The full system shown in Figure 3.16 uses the same base converter simulated in Figure 3.13, copied four times and connected in IPOS configuration. The same large input capacitor and load capacitor on the output are used. The main difference is that the PWM soft start switching profile has been extended to repeat the 33 μs charge time for ~10 ms. Shown in Figure 3.15, the single converter reaches full charge well within the 33 μs window. Thus, the charging window has been decreased to 29 μs. Table 1.1 gives the requirement that the system must repeat the charge cycle up to 300 times, and so a 9 – 10 ms simulation window is sufficient to verify this requirement. To discharge the system after every individual charge cycle, a separate switch was added in parallel to the load capacitor. After the full-bridge converters switch for 29 μs, there is a 2 μs delay where the discharge switch opens and depletes the load capacitor. With the combination of the shortened charge window and the added discharge window, the total system charging frequency still operates at 30 kHz. Figure 3.17 shows the output voltage of the system after one single charge and discharge cycle.

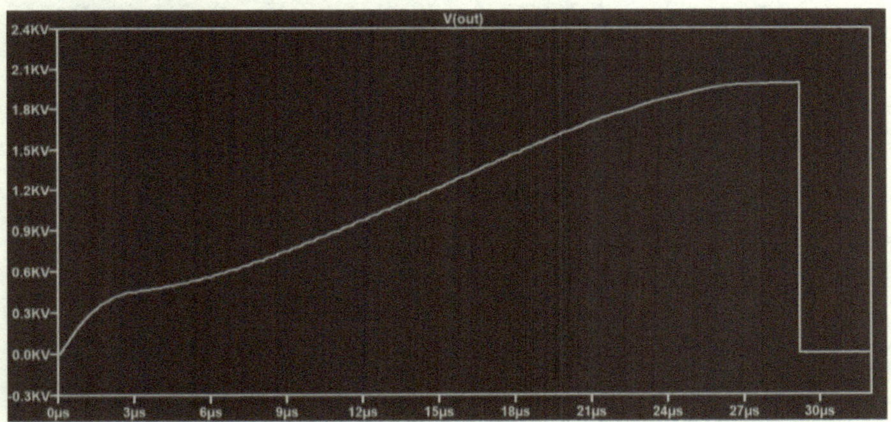

Figure 3.17 Four converter simulation single charge output voltage

As seen above in Figure 3.17, given a 500 V input, the entire simulated four-stack system charges up the output capacitor to ~2 kV in the 29 μs charge window. After the charge time, the load capacitor successfully discharges down to 0 V. The next check that must be done is the currents on the rectifying diodes and output filter inductor, to make sure that it does not exceed the part ratings. Figure 3.18 shows the current on the output of the system as well as a single rectifying diode for a single charge cycle.

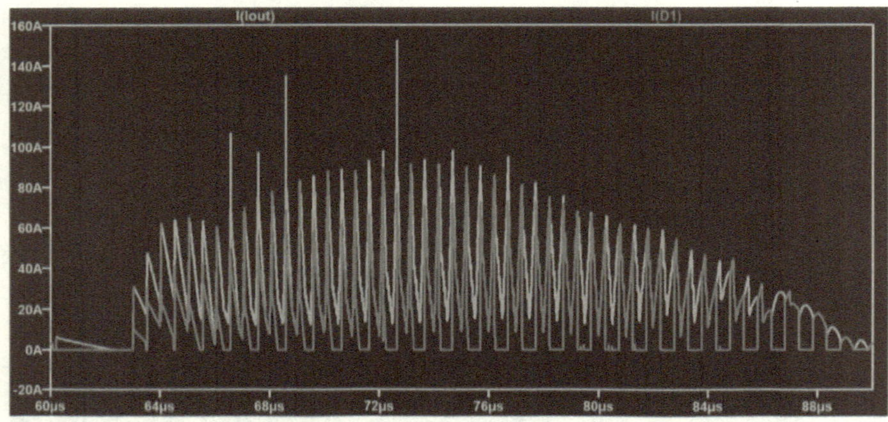

Figure 3.18 Four converter simulation single charge current

In the output shown above, the green waveform corresponds to the system output current, and the red corresponds to the current seen by a single diode. As expected, the current spikes and falls during every switching cycle. For the diode, the peak current value reaches almost 100 A, which is still well beneath the 400 A surge current of the GD10MPS12A part selected in 3.3.5. For the output inductor, the RMS current value must be considered to ensure the inductor does not saturate. Using the math tool in LTSpice, an RMS current of ~ 40 A was found for the output current of a single charge, shown below in Figure 3.19.

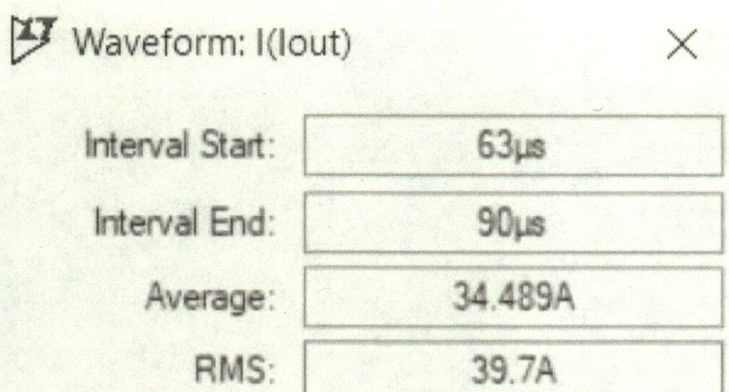

Figure 3.19 LTSpice output current calculation

This 39.7 A RMS is low enough to not saturate the output inductor during operation. The DC saturation current for the selected part is 63 A, as discussed in 3.3.6. With the system functioning as expected in a single charge, a full 300 charge engagement can be simulated. Figure 3.20 shows the first three charge cycles of the 300, and Figure 3.21 shows all 300 consecutive charges. In Figure 3.20, the 29 µs charge time followed by the 2 µs discharge and the 2 µs delay is clearly illustrated, resulting in the system successfully charging to full voltage at 30 kHz.

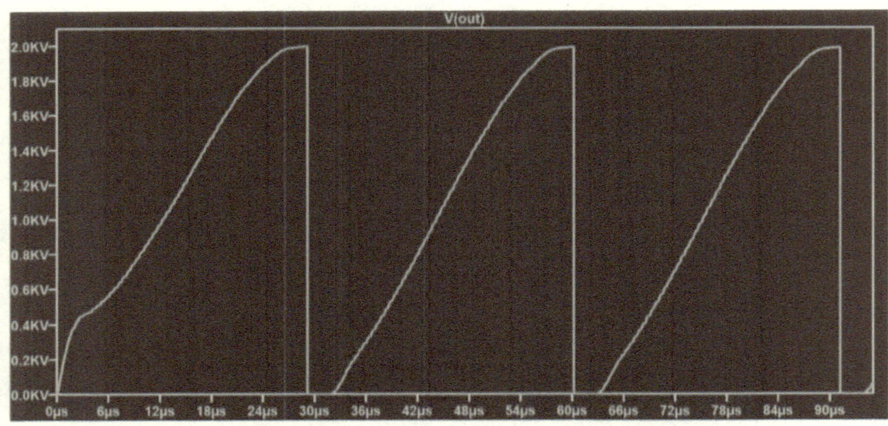

Figure 3.20 Four converter simulation 3 consecutive charges

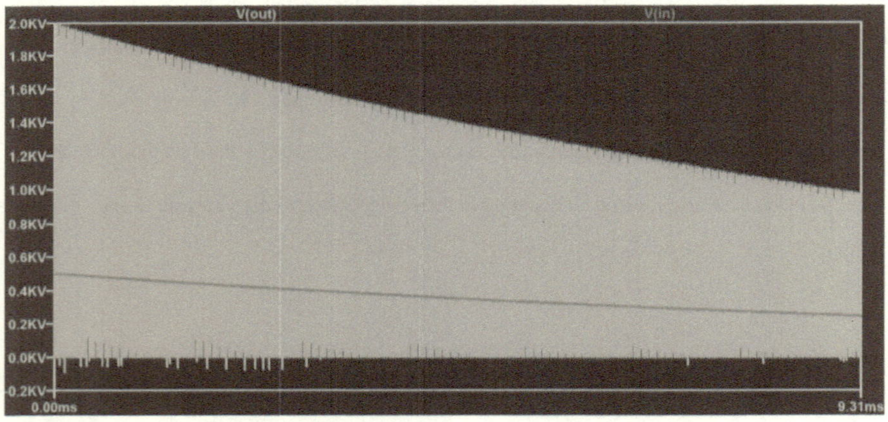

Figure 3.21 Four converter simulation 300 consecutive charges

In Figure 3.21, the 300 consecutive output voltage charge cycles are shown for a 9.5 ms engagement time. While each individual pulse is difficult to discern at the resolution available in LTSpice, zooming in anywhere on the simulation window shows each pulse clearly charging and discharging identically to the clear pulses shown in Figure 3.17 and Figure 3.20. It is noteworthy that over the entire engagement, the input voltage (shown by the red waveform) drops significantly as the input capacitor bank is depleted. As the input voltage drops, the output voltage follows, and the final output voltage pulse is approximately half of the initial pulse. Figure 3.22 illustrates this by comparing the waveform of the last output voltage pulse to the first.

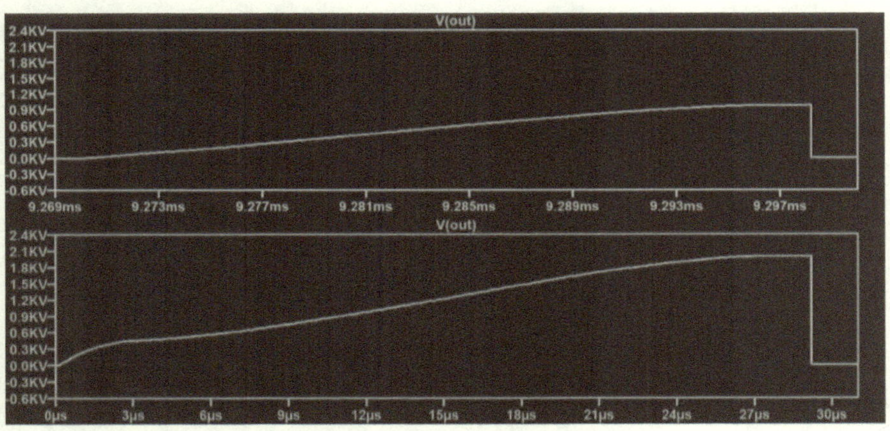

Figure 3.22 Full converter simulation last pulse vs. first pulse

While not a critical system issue, it is important to be aware of this drop in output voltage as the system runs moving into the experimentation phase. With the parts selected and verified by simulation, the system hardware can be produced, as discussed in the next chapter.

CHAPTER IV

EXPERIMENTAL EVALUATION

4.1 Charger Module Board

The full bridge converter discussed and designed in CHAPTER III was successfully implemented in a hardware package, shown in Figure 4.1. The dimensions of a single board are 4.9" long and 2.5" wide, with the tallest components measuring at 0.4" high. Figure 4.2 shows a general circuit diagram of the design with a color code to show the corresponding circuitry in Figure 4.1. In the green section, the PWM control signals enter the board through the pair of gold SMA connectors at the top, with the 5 V power rail being supplied by the connector in between. The GaN MOSFETs, gate driver, and gate drive power supplies are in the bottom right of the green section. It is important to note that all the switching circuitry is contained in a very compact form factor of about one inch by one inch. One half-bridge branch is located on the top of the board, and the other half is mirrored on the bottom. The blue section shows the HV input entering the board through the black connector at the top, with the filter inductors located right beneath it. The two Ceralink filter capacitors are located closer to the switching circuitry to minimize parasitic inductances. The planar transformer, shown in the red box, isolates the input and output sides of the board, and is secured with 4 screws. This allows the transformer to be easily removed and replaced if a different transformer design is desired. Finally, at the bottom of the PCB, shown in orange, are the rectifying diodes, output inductor, and output connector.

Figure 4.1 Front and back of hardware implementation to be tested.

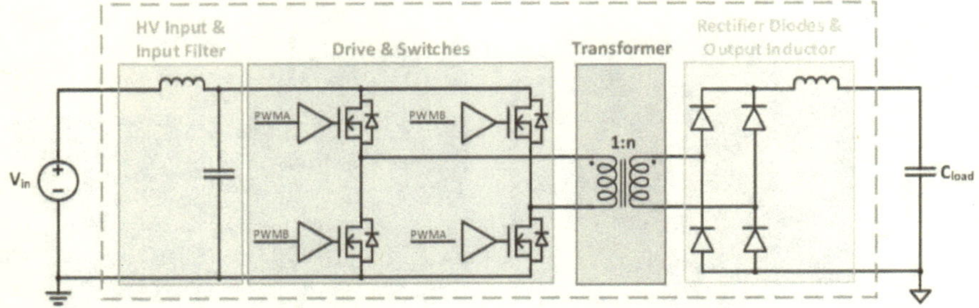

Figure 4.2 Circuit diagram corresponding to hardware implementation.

The diodes are bent down against the board to reduce the overall height, and their thermal pads are facing away from the board to allow for easy cooling, if needed. Two diodes are visible on the front side of the board, and the other two are mirrored identically on the back. The output power leaves the board through the black connector on the very bottom.

4.2 Test Setup

As discussed in CHAPTER I, the board was designed to fit a specific use case at Radiance Technologies. Thus, special test hardware had to be designed to verify the boards' functionality in those conditions. Figure 4.3 shows the complete system setup as designed and tested, and the following sections detail each part shown.

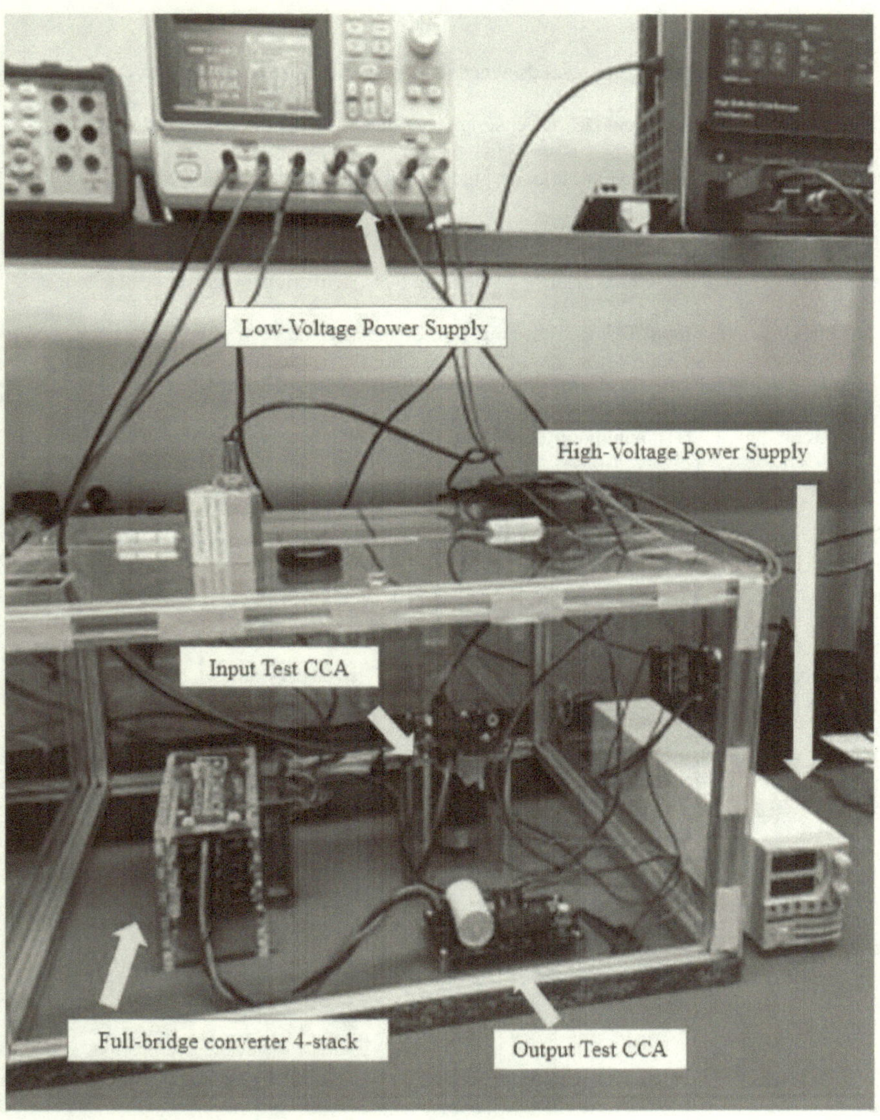

Figure 4.3 Benchtop test setup

4.2.1 Input Test CCA

As discussed, the system was designed to be powered by a capacitor bank, due to its high pulsed power draw. A normal DC power supply could not provide the energy necessary for extended rapid charging and discharging of the system. Thus, a board was designed to simulate the characteristics of a capacitor bank by utilizing a large 1800 μF electrolytic capacitor. This board hooked up to the input connector of the main PCB, delivering the power needed to test. The circuitry on the Input Test CCA is simple, receiving an input from a high-voltage DC power supply to charge the capacitor, with a dump circuit to bleed the capacitor when testing is done. The circuit diagram is shown in Figure 4.4, and the assembled hardware is shown in Figure 4.5.

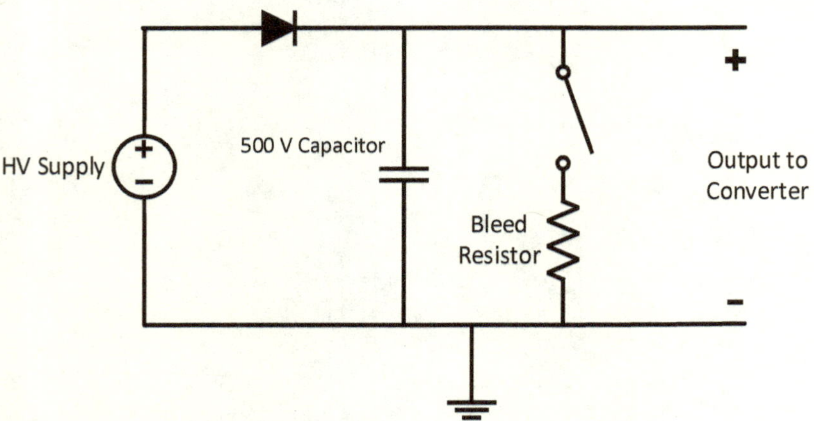

Figure 4.4 Input Test CCA simplified circuit diagram

The assembled hardware can charge up to 500 V and discharging into the full bridge converter during operation. When testing is complete, the capacitor safely discharges with a flick of a switch using the large bleed resistor and relay.

Figure 4.5 Input Test CCA hardware

4.2.2 Output Test CCA

The modular charger system was designed to charge up a specialized spiral generator that has characteristics like a capacitor, as discussed in 1.2. To ensure the system meets the design requirements, an Output Test CCA was designed to simulate this spiral generator charging and discharging at a defined pulse repetition frequency. The circuitry of this peripheral CCA is shown by Figure 4.6.

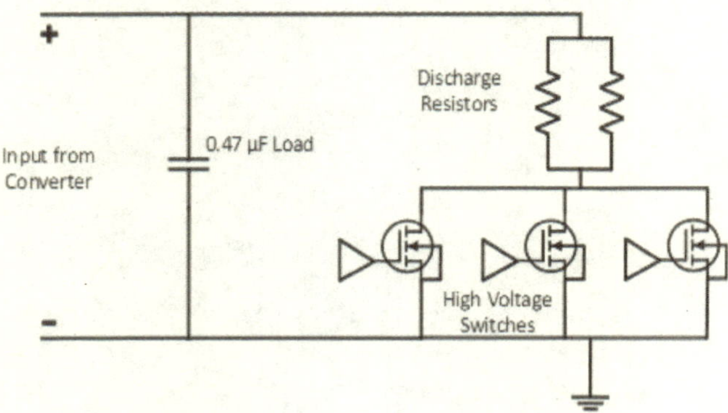

Figure 4.6 Output Test CCA circuit diagram

The full-bridge converter system outputs directly into a 0.47 µF capacitor on the Output Test CCA. This capacitance value simulates the spiral generator this system was designed for. The capacitor can be rapidly discharged through two 1 Ω Ohmite PulsEater resistors connected in parallel [60]. These specialized resistors can quickly absorb large amounts of pulsed energy, allowing for the rapid discharge of the load capacitor. The discharge signal is provided to the board externally via BNC connector, allowing for precise control of the discharge PRF. 3 SiC MOSFETs serve as the discharge switches. These parts were selected due to their high voltage and current ratings and low R_{DSon} value, allowing them to handle the rapid discharge of high energy through them. The completed hardware for the Output Test CCA is shown in Figure 4.7.

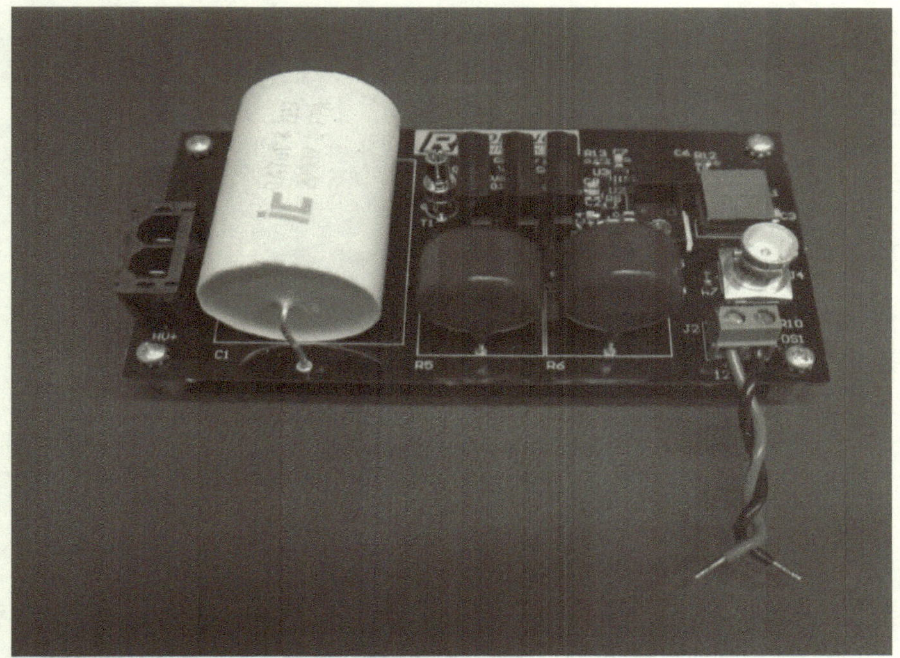

Figure 4.7 Output Test CCA hardware

4.2.3 FPGA

To drive the main full-bridge converter and the peripheral test hardware, a configurable solution is required. The ability to change duty cycles and adjust PRF and charge times is necessary to fully test the capabilities of the whole system. To meet these needs, a Lattice FPGA development board was equipped with BNC breakout boards to connect to up to 8 full-bridge converter PCBs as well as the discharge signal on the Output Test CCA, shown in Figure 4.8.

Figure 4.8 Control FPGA with BNC breakout board

A GUI was created by another team member to run on this board, allowing for interface between a laptop and the FPGA. This GUI, shown in Figure 4.9, allowed for fine configuration of each charge, with the ability to set the switching period of the charge, the number of pulses in a charge, the pulse width of the discharge signal, the number of charges in an engagement, and the number of total engagements. The ability to configure each charge in such detail became a vital part of the testing process.

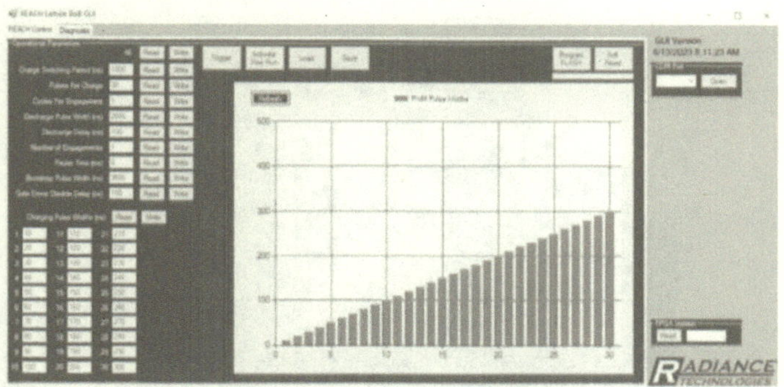

Figure 4.9 System control interface

4.2.4 Modular Stack Setup

Because the four individual converters are connected in an IPOS configuration as discussed in 1.2.3, a special hardware implementation is needed. Also, the restricted form factor necessitates the four boards be mounted as tightly as possible. Thus, an acrylic enclosure was designed with card slots for each converter to slide in to. Figure 4.10 shows an initial CAD rendering of the assembly, and Figure 4.11 shows the final hardware. The completed hardware allows for easy access to each boards' input connectors. The input power is paralleled from the Input Test CCA in 4.2.1 via terminal block and wired into the high voltage connector on each card. Each PWM signal is taken from the FPGA in 4.2.3 into the pair of SMA connectors on the converters. The 5 V power rail is supplied into each PCB by a benchtop power supply. On the output side, a special wire harness was created to provide the series connection.

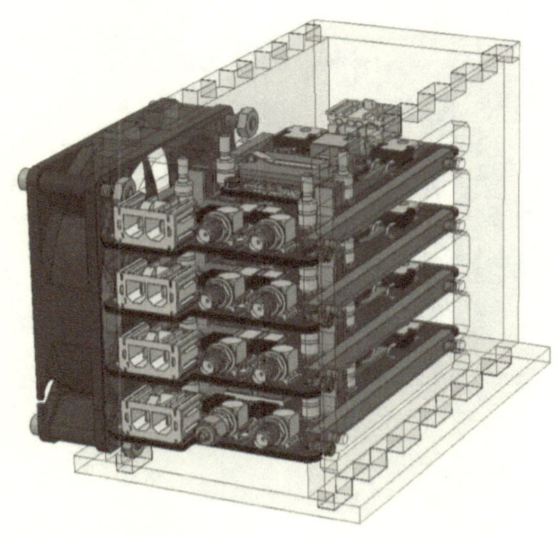

Figure 4.10 IPOS converter assembly 3D model

Figure 4.11 IPOS converter assembly finished hardware

4.3 Results

4.3.1 Single Card Testing

Before testing the system in its full IPOS configuration with four converters, an evaluation of a single converter was needed. To begin evaluation, a single card was connected only to the FPGA and 5 V power supply to test the gate voltages of the GaN switches, and whether the PWM signals were communicating with the gate driver properly. Figure 4.12 shows the results of this low-voltage test.

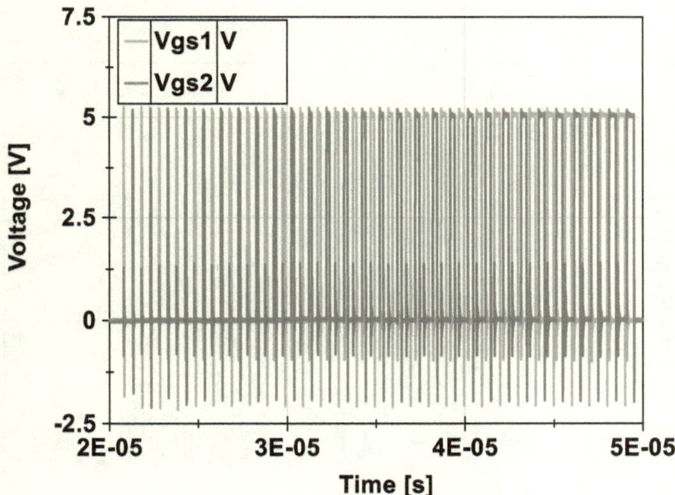

Figure 4.12 Gate voltage test of single converter

The complimentary nature of the high-side and low-side gate signals can clearly be seen in Figure 4.12, as they successfully switch at 5 V. The "soft-start" switching profile can also be seen, as the pulse widths of the signals widen with each switch, reaching the full 60% duty cycle by the end of the charge time. The 1 μs switching period successfully demonstrates the GaN HEMTs switching at a 1 MHz switching frequency. With proper communication between the FPGA, gate driver, and switches, an input voltage could be given to the system. The input test CCA discussed in 4.2.1. was connected to a high-voltage power supply and to the input of the full-bridge converter. The planar transformer was disconnected to decouple the converter input from the output, and 20 V was applied to the system by the power supply. Figure 4.13 shows the result of this test, with measurements taken at the outputs of the switches into the transformer nodes.

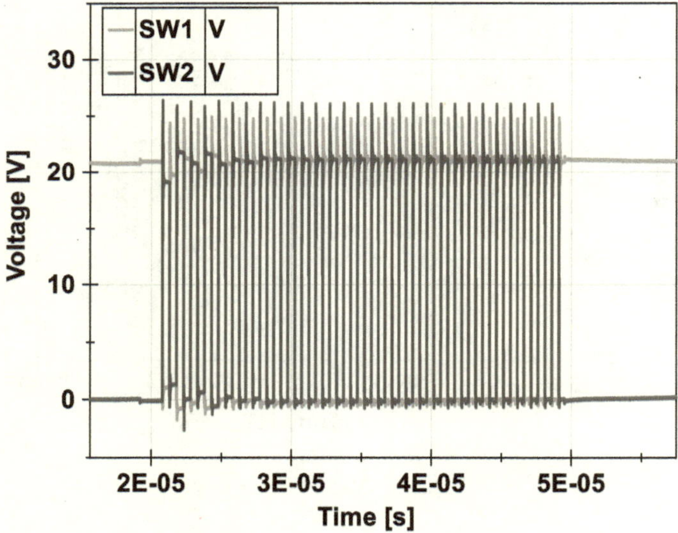

Figure 4.13 Switch outputs with 20 V input

As shown above, the GaN switches successfully operated at the 20 V level. Some ground bounce was seen at the beginning of the waveform, most likely due to the lack of an inductive load on the switches. But with verified operation that all the switching circuitry was functioning correctly, more voltage could be applied to the system. The planar transformer and output test CCA were both connected to supply the proper 0.47 μF output load to the system. Once the input voltage was incremented, an issue started to arise with the system output. The output voltage experienced dips during the charge window and thus was not reaching the full voltage charge level. Also, the voltage on the output was not being discharged correctly on the output test CCA. Figure 4.14 illustrates this phenomenon.

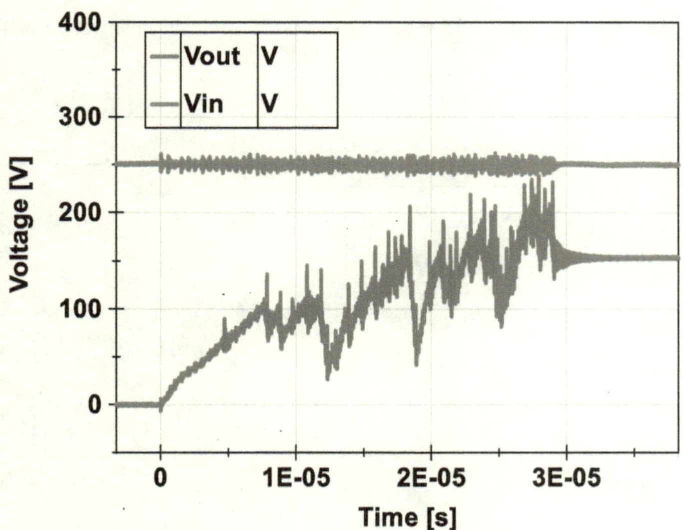

Figure 4.14 Output voltage issue

On the specific test shown in Figure 4.14, a 250 V input was applied to the converter (shown in red). With the 1:1 conversion factor on the transformer, a final output voltage level of 250 V was expected. But clearly, the output voltage waveform shown in green experienced a significant amount of noise and voltage drop through the charge window. A smooth output voltage curve like the one simulated in Figure 4.6 on page 45 should have been seen. After debugging the system, an issue was found on the discharge signal coming from the control FPGA to the output test CCA. One of the connections was not soldered properly, and thus the discharge signal was left floating. This floating signal allowed for system noise on the discharge switches on the output test CCA (Figure 4.6 on page 58 for reference). Thus, it erroneously discharged during the charge cycle, and did not correctly discharge when the charge cycle was complete. Once the connection on the discharge signal was restored, the system began charging and discharging properly. Figure 4.15 shows the successful charging and discharging of the system's output voltage.

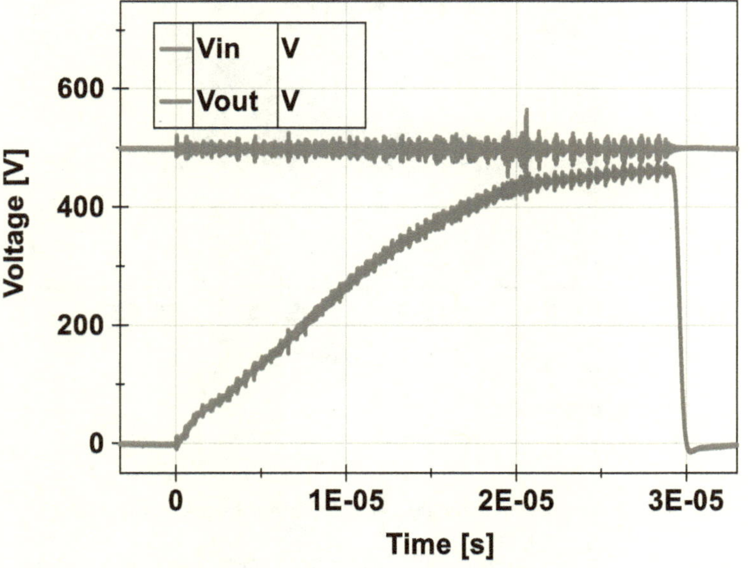

Figure 4.15 Single card 500 V input successful charge

Given a 500 V input, the single card charged the 0.47 μF load capacitor to 460 V before rapidly discharging. While not reaching an exact 1:1 voltage conversion, the system outputted a voltage level of 92% of the input, which is acceptable considering the various loss elements at play in the PCB. This entire charge happened in 29 μs, allowing 4 μs of discharge and idle before having to charge again to meet the required 30 kHz PRF. This is further demonstrated in Figure 4.16, where the converter is consecutively charged and discharged twice, showing a successful 30 kHz PRF. The red input waveform in each graph experiences noise during the duration of each charge cycle, but the maintaining of a constant voltage level shows a successful integration of the system's input filter.

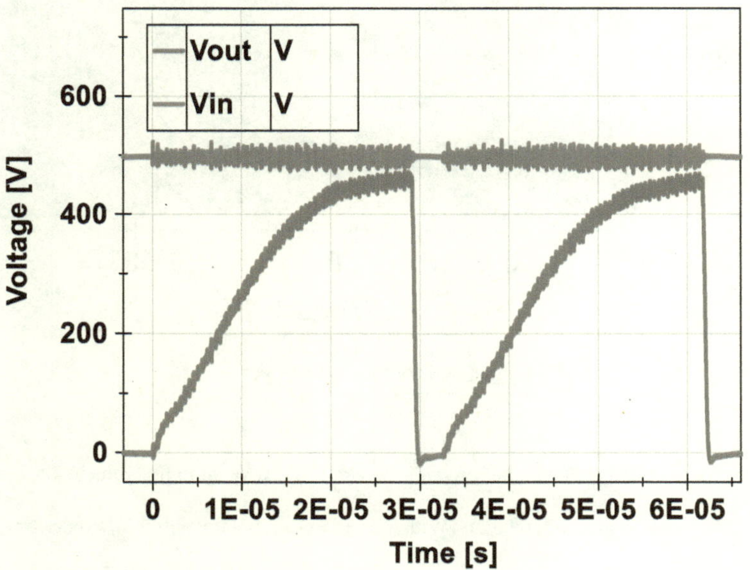

Figure 4.16 Single card 500 V input successful consecutive charges

Finally, Figure 4.17 shows the single converter operating for a full 300 charge engagement, meaning the converter charged and discharged the load capacitor 300 consecutive times at a 30 kHz PRF.

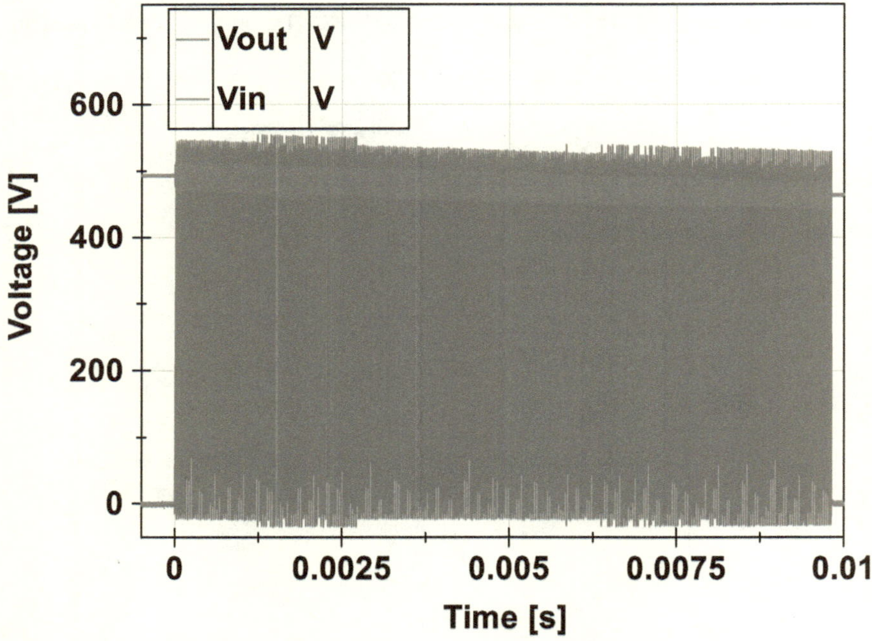

Figure 4.17 Single card 500 V input 300 consecutive charges

As demonstrated above, the converter was able to charge up to full voltage 300 consecutive times, in just under 9.3 ms. With the successful functionality of a single converter PCB verified, the system is ready to be connected and evaluated in the full four converter IPOS configuration.

4.3.2 Four Card Testing

Once the four converters were connected in the IPOS configuration shown in Figure 4.11 on page 62, there was no way to effectively probe the individual switch gate signals to verify system functionality before applying an input voltage. But since each card had been verified to work correctly in the last section, it was decided to proceed in applying input voltage to the entire system. The input voltage was incremented up to 450 V with no issues; Figure 4.18 shows the result of this test.

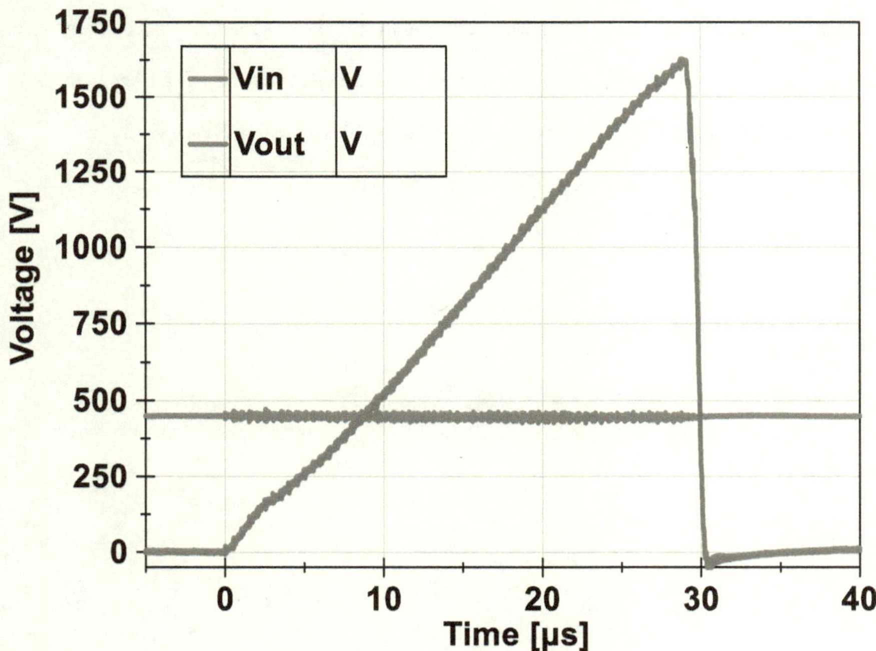

Figure 4.18 Four converter IPOS configuration 450 V input

As seen above, the IPOS configuration works as expected, and was able to multiply the input voltage. With a 450 V input, the expected output voltage in an ideal converter would be 1800 V, due to the four converters each outputting 450 V, and then stacking together in series. Figure 4.18 shows that the total system output voltage is 1625 V, which is not quite the ideal 1800 V. There are two factors that affect this. First, there are physical losses in the hardware due to heat and parasitics, as was seen in the single converter testing. Also, given a larger time scale, the system could have continued to charge until it hit a higher voltage level. But with the 30 kHz PRF requirement, the load must discharge at 29 μs to allow for discharge and idle times before charging back up. Despite these two factors, the system still exceeds requirements by charging well above 1200 V in under the 33 μs timeframe, which is a great success. To further evaluate how the IPOS configuration functions, Figure 4.19 shows the current into each of the four converters, with the total current out of the large input capacitor.

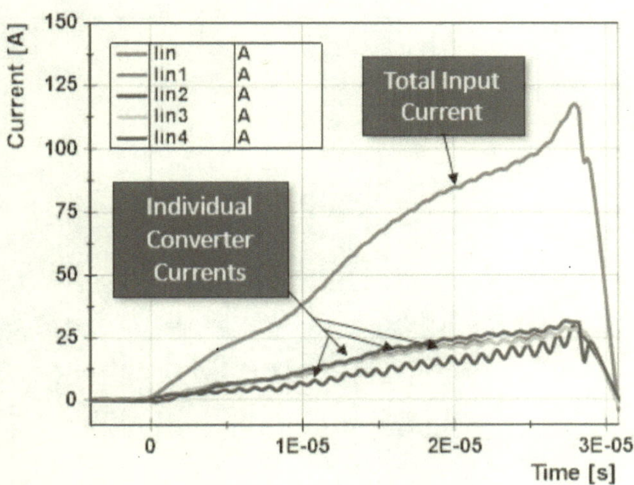

Figure 4.19 Four converter IPOS configuration input current sharing

The total input current (shown in red) is distributed evenly across the four converters, with only an approximately 10% mismatch between any two cards. Figure 4.18 and Figure 4.19 in conjunction show the efficacy of the IPOS configuration, namely the current sharing across the input of each converter and the voltage multiplying across the output. To calculate the peak output power of the system, the output current must be measured. Figure 4.20 shows the total output current of the system.

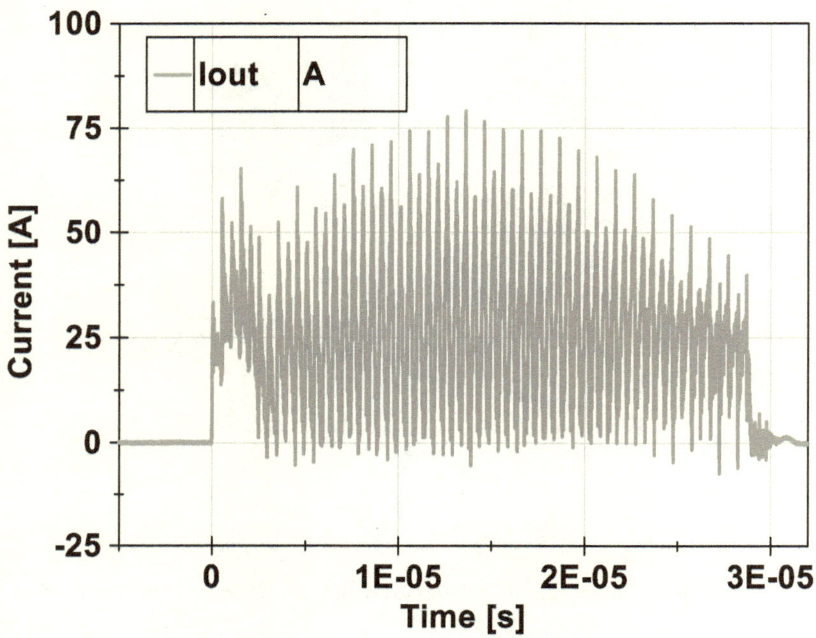

Figure 4.20 Four converter IPOS configuration output current

As seen above, the peak output current of the system is 79 A. When the output current waveform in Figure 4.20 is combined with the output voltage waveform in Figure 4.18, the instantaneous peak output power can be calculated. The highest peak power the system achieves at a given point during the charge is 81.5 kW, occurring at 22.67 μs into the charge cycle. This value is calculated by multiplying the output current peak by the output voltage at that instant. Due to the nature of this pulsed power device, the instantaneous peak power is simple to calculate, but the average power is more difficult. Because the traditional way to calculate power, multiplying voltage by current, changes at any given point because of the pulsed current values, the load must be looked at to calculate average power delivery. Since the load is capacitive, the equation for energy in a capacitor can be used, shown in Equation 4.1.

$$E = \frac{1}{2}CV^2 \tag{4.1}$$

When substituting the relationship between power and energy, shown earlier in Equation 1.4, into Equation 4.1, the following relationship appears.

$$P = \frac{CV^2}{2t} \tag{4.2}$$

Equation 4.2 can be used to calculate the average output power of the system, where C is the load capacitance of 0.47 μF, V is the maximum output voltage of 1625 V, and t is the charge time of 29 μs. Plugging these numbers into Equation 4.2 yields the average output power a single charge: 21.4 kW. This value exceeds the required output power of 10.25 kW given in Table 1.1. Finally, the last test to show successful fulfillment of the system requirements is to run consecutive charges on the IPOS configuration. Figure 4.21 shows the entire system running for 300 consecutive charges, as required.

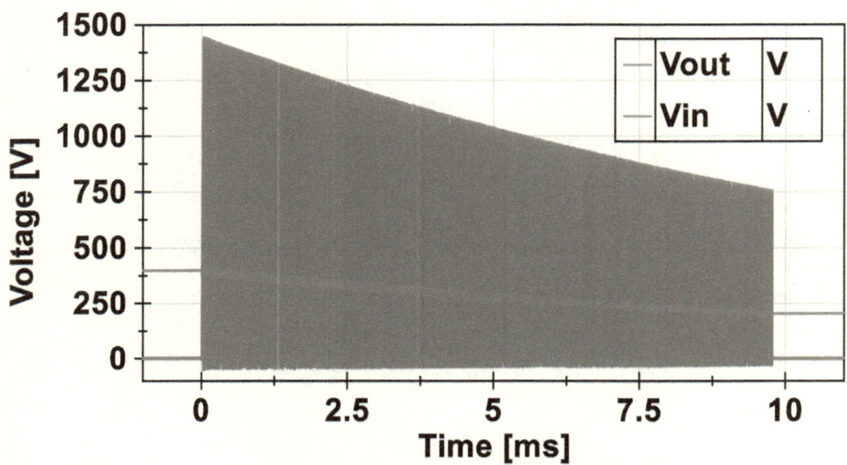

Figure 4.21 Four converter IPOS configuration 300 consecutive charges

Similar to the IPOS configuration simulation results in Figure 3.21 on page 50, the system experiences significant input voltage droop, resulting in an output voltage drop between the first charge and last charge. This can be fixed by increasing the size of the capacitor bank on the input of the system. Despite the voltage drop across the engagement, the system successfully charges and discharges to an average voltage > 1200 V 300 consecutive times. Finally, the requirements in Table 1.1 state that the system must be able to repeat the 300 consecutive charge engagement given a 1 – 5 second idle time. Figure 4.22 shows the successful fulfillment of that requirement, with two 300 charge engagements shown with a 1 second idle team in between.

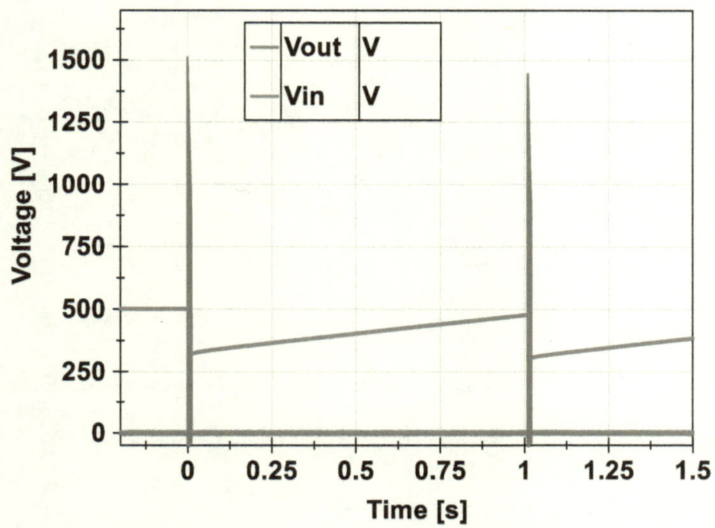

Figure 4.22 Four converter IPOS configuration two 300 charge engagements

The 1 second idle time after the first 300 charge engagement is enough to allow the voltage on the input capacitor to almost recharge fully before starting the next 300 charge engagement. As mentioned earlier, a larger capacitor bank and a large power supply would mitigate the voltage drop across engagements, but this is a factor of the peripheral test setup, and not the converter system itself. Thus, the system is verified to meet the requirements laid out at the beginning of this book.

CHAPTER V

CONCLUSION & FUTURE WORK

5.1 Conclusion

To summarize the overall system performance, Table 5.1 compares the required values given in Table 1.1 to the experimentally verified system values shown in 3.3.5.

Table 5.1 System Requirements vs. Experimental Results

Design Parameter	Required Value	Verified Value
Input Voltage	400 - 500 V	400 - 500 V
Output Voltage	≥ 1200 V	1625 V
Load Capacitance	0.47 μF	0.47 μF
Charge Time	< 33 μs	29 μs
Average Power	≥ 10.25 kW	21.4 kW
Pulse Repetition Frequency (PRF)	≥ 30 kHz	30 kHz
Switching Frequency	1 MHz	1 MHz
# of Consecutive Capacitor Charges	300	300
Idle/Recharge Time	1-5 s	1 s
Initial Size	4 lbs., 80 in^3	1.5 lbs., 60 in^3

As illustrated in Table 5.1 above, each system requirement for the design in this book has been satisfied or exceeded. The system was shown to take an input up to 500 V, and output 1600 V in a 29 μs charge time, corresponding to an average output power of 21.4 kW and a peak power of 64.3 kW. 300 consecutive charges at full voltage were accomplished at a 30 kHz PRF, while the individual full-bridge GaN transistors switched at 1 MHz. The system was also able to perform multiple consecutive 300 charge engagements with a 1 second idle time in between for the input capacitor to recharge. And finally, the assembled system was lighter and smaller than the already compact required form factor of 4 lbs., 80 in^3, coming in at 1.5 lbs., 60 in^3. In conclusion, regarding the requirements this design set out to meet, the work in this book is a success.

5.2 Future Work

Despite the successful fulfillment of requirements, there is still future work that can be done to further improve or adapt this system. When compared to the pulsed power capacitor chargers on the market discussed in CHAPTER II, the design in this book reaches PRFs significantly faster than any existing system, in a much smaller form factor. This level of portability opens opportunities for integration with many different fielded HPM and pulsed power systems. While work is actively being done integrating this specific system with other parts of an HPM device at Radiance Technologies in Huntsville, AL, the flexibility and adaptability of the system allows for customization beyond this specific HPM device. The scalable IPOS configuration allows for more than four individual converters to be connected, allowing for even higher achievable output power levels. With the easy drop-in form factor of the planar transformer, different transformer turns ratios can be used, either to step up or step down the output voltages. Both customization options can be manipulated in unison to allow for tunability to a wide range of output voltage and power requirements. These features all come by nature of the modular design of this project and open many avenues for ongoing experimentation and future work.

REFERENCES

[1] T. H. Martin, A. H. Guenther, and M. Kristiansen, Eds., *J. C. Martin on Pulsed Power*. Boston, MA: Springer US, 1996. doi: 10.1007/978-1-4899-1561-0.

[2] S.-H. Kim, M. Ehsani, and C.-S. Kim, "High-voltage power supply using series-connected full-bridge PWM converter for pulsed power applications," *IEEE Trans. Dielect. Electr. Insul.*, vol. 22, no. 4, pp. 1937–1944, Aug. 2015, doi: 10.1109/TDEI.2015.004978.

[3] J. Zaiter, "An Overview of Pulsed Power System Design and Applications," book, 2021

[4] F. Davanloo, C. B. Collins, and F. J. Agee, "Development and applications of pulsed power devices at the University of Texas at Dallas," in *Digest of Technical Papers. PPC-2003. 14th IEEE International Pulsed Power Conference (IEEE Cat. No.03CH37472)*, Dallas, TX, USA: IEEE, 2003, pp. 1403–1406. doi: 10.1109/PPC.2003.1278078.

[5] H. Akiyama, T. Sakugawa, T. Namihira, K. Takaki, and Y. Minamitani, "Industrial Applications of Pulsed Power Technology," vol. 14, no. 5, 2007.

[6] K. M. Sayler, A. Feickert, and R. O'Rourke, "Department of Defense Directed Energy Weapons: Background and Issues for Congress," Congressional Research Service, 2023.

[7] S. Gutierrez and S. Beeson, "Introduction to High Power Microwave Systems: HPM Fundamentals," Kirtland AFB, NM 87117, 2019.

[8] S. T. Pai and Q. Zhang, *Introduction to High Power Pulse Technology*, vol. 10. World Scientific, 1995.

[9] G. H. Rim *et al.*, "A constant current high voltage capacitor charging power supply for pulsed power applications," in *PPPS-2001 Pulsed Power Plasma Science 2001. 28th IEEE International Conference on Plasma Science and 13th IEEE International Pulsed Power Conference. Digest of Papers (Cat. No.01CH37251)*, Las Vegas, NV, USA: IEEE, 2001, pp. 1284–1286 vol.2. doi: 10.1109/PPPS.2001.1001783.

[10] M. Giesselmann, T. Heeren, and T. Helle, "Compact, high power capacitor charger," in *Digest of Technical Papers. PPC-2003. 14th IEEE International Pulsed Power Conference (IEEE Cat. No.03CH37472)*, Dallas, Texas, USA: IEEE, 2003, pp. 707–710. doi: 10.1109/PPC.2003.1277806.

[11] A. Bole, *Radar and ARPA Manual Radar and Target Tracking for Professional Mariners, Yachtsmen and Users of Marine Radar*, 3rd ed. 2014.

[12] I. J. Cohen, E. Glynn, A. King, Z. Roberts, and M. Heffernan, "Evaluation of Liquid Dielectrics to Improve Spiral Generator Lifetimes," in *2021 IEEE Pulsed Power Conference (PPC)*, Denver, CO, USA: IEEE, Dec. 2021, pp. 1–7. doi: 10.1109/PPC40517.2021.9733136.

[13] "An Introduction to Switch-Mode Power Supplies." Maxim Integrated, Sep. 27, 2007.

[14] M. Zehendner and M. Ulmann, "Power Topologies Handbook." Texas Instruments.

[15] R. Erickson and D. Maksimovic, *Fundamentals of Power Electronics*, 2nd ed. Springer Science & Business Media, 2001.

[16] Y. Zhang, X. Li, Z. Miu, K. Kundanam, J. Liu, and Y. Liu, "High step-up full bridge DC-DC converter with multi-cell diode-capacitor network," in *2017 IEEE Applied Power Electronics Conference and Exposition (APEC)*, Tampa, FL, USA: IEEE, Mar. 2017, pp. 157–163. doi: 10.1109/APEC.2017.7930687.

[17] W. Chen, X. Ruan, H. Yan, and C. K. Tse, "DC/DC Conversion Systems Consisting of Multiple Converter Modules: Stability, Control, and Experimental Verifications," *IEEE Trans. Power Electron.*, vol. 24, no. 6, pp. 1463–1474, Jun. 2009, doi: 10.1109/TPEL.2009.2012406.

[18] J. You, L. Cheng, B. Fu, and M. Deng, "Analysis and Control of Input-Parallel Output-Series Based Combined DC/DC Converter With Modified Connection in Output Filter Circuit," *IEEE Access*, vol. 7, pp. 58264–58276, 2019, doi: 10.1109/ACCESS.2019.2914558.

[19] J. Millan, P. Godignon, X. Perpina, A. Perez-Tomas, and J. Rebollo, "A Survey of Wide Bandgap Power Semiconductor Devices," *IEEE Trans. Power Electron.*, vol. 29, no. 5, pp. 2155–2163, May 2014, doi: 10.1109/TPEL.2013.2268900.

[20] A. Lidow, J. Strydom, M. D. Rooij, and D. Reusch, *GaN Transistors for Efficient Power Conversion*, 2nd ed. John Wily & Sons Ltd, 2015.

[21] "An Introduction to GaN Enhancement-mode HEMTs." GaN Systems, Inc., Mar. 08, 2022.

[22] P. Thummala, D. B. Yelaverthi, R. A. Zane, Z. Ouyang, and M. A. E. Andersen, "A 10-MHz GaNFET-Based-Isolated High Step-Down DC–DC Converter: Design and Magnetics Investigation," *IEEE Trans. on Ind. Applicat.*, vol. 55, no. 4, pp. 3889–3900, Jul. 2019, doi: 10.1109/TIA.2019.2904455.

[23] Y. Yao, G. S. Kulothungan, H. S. Krishnamoorthy, A. Das, and H. Soni, "GaN-Based Two-Stage Converter With High Power Density and Fast Response for Pulsed Load Applications," *IEEE Trans. Ind. Electron.*, vol. 69, no. 10, pp. 10035–10044, Oct. 2022, doi: 10.1109/TIE.2022.3159946.

[24] L. T. Alex, V. Jaikrishna, S. S. Dash, and R. Sridhar, "Design and analysis of Push-pull-Flyback interleaved converters for photovoltaic system," in *2017 IEEE 6th International Conference on Renewable Energy Research and Applications (ICRERA)*, San Diego, CA: IEEE, Nov. 2017, pp. 757–761. doi: 10.1109/ICRERA.2017.8191161.

[25] C. Wang, M. Li, Z. Ouyang, and G. Wang, "High Efficiency High Step-up Isolated DC-DC Converter for Photovoltaic Applications," in *2019 IEEE Applied Power Electronics Conference and Exposition (APEC)*, Anaheim, CA, USA: IEEE, Mar. 2019, pp. 1273–1280. doi: 10.1109/APEC.2019.8721916.

[26] V. F. Gruner, L. Fiamoncini, L. Schmitz, D. C. Martins, and R. Francisco Coelho, "High Step-Up DC-DC Converter with Input Current Sharing Based on the Forward Converter," in *2019 IEEE 15th Brazilian Power Electronics Conference and 5th IEEE Southern Power Electronics Conference (COBEP/SPEC)*, Santos, Brazil: IEEE, Dec. 2019, pp. 1–6. doi: 10.1109/COBEP/SPEC44138.2019.9065676.

[27] L. Wang, "Input-Parallel and Output-Series Modular DC-DC Converters with One Common Filter," in *EUROCON 2007 - The International Conference on "Computer as a Tool,"* Warsaw, Poland: IEEE, 2007, pp. 1398–1402. doi: 10.1109/EURCON.2007.4400427.

[28] Yue Zhang, Zheng Wang, and Ming Cheng, "An interleaved current-fed bidirectional full-bridge DC/DC converter for on-board charger," in *IECON 2016 - 42nd Annual Conference of the IEEE Industrial Electronics Society*, Florence, Italy: IEEE, Oct. 2016, pp. 4376–4381. doi: 10.1109/IECON.2016.7793532.

[29] Y. Zhang, Z. Wang, M. Cheng, and L. Xu, "Input-Parallel Output-Series DC/AC Converter for On-Board EV Charger," in *2016 IEEE Vehicle Power and Propulsion Conference (VPPC)*, Hangzhou, China: IEEE, Oct. 2016, pp. 1–5. doi: 10.1109/VPPC.2016.7791710.

[30] Z. Yuan, H. Xu, Y. Chao, and Z. Zhang, "A novel fast charging system for electrical vehicles based on input-parallel output-parallel and output-series," in *2017 IEEE Transportation Electrification Conference and Expo, Asia-Pacific (ITEC Asia-Pacific)*, Harbin, China: IEEE, Aug. 2017, pp. 1–6. doi: 10.1109/ITEC-AP.2017.8080843.

[31] Zhejiang University and Y. Li, "High-Gain High-Efficiency IPOS LLC Converter With Coupled Transformer and Current Sharing Capability," *CPSS TPEA*, vol. 5, no. 1, pp. 63–73, Mar. 2020, doi: 10.24295/CPSSTPEA.2020.00006.

[32] Chung-Ming Young, Jhih-Wun Siao, Wei-Shan Yeh, and Shih-Jen Cheng, "An input-parallel and output-series-parallel phase-shift full-bridge converter with maximum power point tracking for wind turbine," in *2013 International Conference on Renewable Energy Research and Applications (ICRERA)*, Madrid, Spain: IEEE, Oct. 2013, pp. 133–136. doi: 10.1109/ICRERA.2013.6749739.

[33] M. Guan, "A Series-Connected Offshore Wind Farm Based on Modular Dual-Active-Bridge (DAB) Isolated DC–DC Converter," *IEEE Trans. Energy Convers.*, vol. 34, no. 3, pp. 1422–1431, Sep. 2019, doi: 10.1109/TEC.2019.2918200.

[34] H. A. Hussain and K. A. N. Al-Deen, "Design and Control of Series-DC Wind Farms based on Three-Phase Dual Active Bridge Converters," in *2022 IEEE Energy Conversion Congress and Exposition (ECCE)*, Detroit, MI, USA: IEEE, Oct. 2022, pp. 1–8. doi: 10.1109/ECCE50734.2022.9947753.

[35] S. Jin, C. Zhang, Y. Peng, and Z. Fang, "Novel IPOx Architecture for High-Voltage Microsecond Pulse Power Supply Using Energy Efficiency and Stability Model Design Method," *IEEE Trans. Power Electron.*, vol. 36, no. 9, pp. 10852–10865, Sep. 2021, doi: 10.1109/TPEL.2021.3064957.

[36] J. Klüss and W. Larzelere, "Reconfiguration of 3 MV Marx Generator into a Modern High Efficiency System," *NORD-IS*, no. 25, Oct. 2017, doi: 10.5324/nordis.v0i25.2372.

[37] S. L. Holt, C. F. Lynn, J. M. Parson, J. C. Dickens, A. A. Neuber, and J. J. Mankowski, "Burst mode operation of a high peak power high pulse repetition rate capacitor charging power supply," in *2015 IEEE Pulsed Power Conference (PPC)*, Austin, TX, USA: IEEE, May 2015, pp. 1–4. doi: 10.1109/PPC.2015.7297029.

[38] "FF600R06ME3 Datasheet." Infineon, Nov. 04, 2013.

[39] Y. Zhang *et al.*, "Design of compact high-voltage capacitor charging power supply for pulsed power application," in *2015 IEEE Pulsed Power Conference (PPC)*, Austin, TX, USA: IEEE, May 2015, pp. 1–5. doi: 10.1109/PPC.2015.7297001.

[40] A. Pokryvailo, C. Carp, and C. Scapellati, "High-Power High-Performance Low-Cost Capacitor Charger Concept and Implementation," *IEEE Trans. Plasma Sci.*, vol. 38, no. 10, pp. 2734–2745, Oct. 2010, doi: 10.1109/TPS.2010.2051959.

[41] S. R. Jang, S. H. Ahn, H. J. Ryoo, and G. H. Rim, "Novel high voltage capacitor charger for pulsed power modulator," in *2010 IEEE International Power Modulator and High Voltage Conference*, Atlanta, GA, USA: IEEE, May 2010, pp. 317–321. doi: 10.1109/IPMHVC.2010.5958357.

[42] M. Giesselmann and E. Kristiansen, "Compact design of a 30 kV rapid capacitor charger," in *PPPS-2001 Pulsed Power Plasma Science 2001. 28th IEEE International Conference on Plasma Science and 13th IEEE International Pulsed Power Conference. Digest of Papers (Cat. No.01CH37251)*, Las Vegas, NV, USA: IEEE, 2001, pp. 640–643. doi: 10.1109/PPPS.2001.1002177.

[43] M. Giesselmann and J. Mayes, "Rapid Capacitor Charger with Advanced Digital Control," in *2021 IEEE Pulsed Power Conference (PPC)*, Denver, CO, USA: IEEE, Dec. 2021, pp. 1–3. doi: 10.1109/PPC40517.2021.9733157.

[44] M. G. Giesselmann, T. T. Vollmer, and L. Altgilbers, "Modular, compact HV-capacitor charger," in *2010 IEEE International Power Modulator and High Voltage Conference*, Atlanta, GA, USA: IEEE, May 2010, pp. 409–412. doi: 10.1109/IPMHVC.2010.5958381.

[45] A. Agarwal, Y. Prabowo, and S. Bhattacharya, "Analysis and Design Considerations of Input Parallel Output Series-Phase Shifted Full Bridge Converter for a High-Voltage Capacitor Charging Power Supply," in *2021 IEEE 12th Energy Conversion Congress & Exposition - Asia (ECCE-Asia)*, Singapore, Singapore: IEEE, May 2021, pp. 1068–1075. doi: 10.1109/ECCE-Asia49820.2021.9479247.

[46] "Phase-Shifted Full Bridge DC/DC Power Converter Design Guide." Texas Instruments, 2014.

[47] J. Dyer *et al.*, "Design and Development of Compact High Repetition Rate kW-Class Capacitor Charger," in *2023 IEEE Pulsed Power Conference (PPC)*, San Antonio, TX, USA: IEEE, Jun. 2023, pp. 1–5. doi: 10.1109/PPC47928.2023.10310919.

[48] "GS66516B Bottom-side cooled 650 V E-mode GaN transistor Datasheet." GaN Systems, Inc., 2021.

[49] "GS66516T Top-side cooled 650 V E-mode GaN transistor Datasheet." GaN Systems, Inc., 2021.

[50] D. Dankov and P. Marinov, "Study of Power GaN MOSFET Gate Drivers," in *2022 13th National Conference with International Participation (ELECTRONICA)*, Sofia, Bulgaria: IEEE, May 2022, pp. 1–4. doi: 10.1109/ELECTRONICA55578.2022.9874434.

[51] "EiceDRIVER 2EDI product family." Infineon, 2022.

[52] "1769205132 MagI3C Power Module WPME-FIMM - Fixed Isolated Micromodule." Wurth Elektronik, 2023.

[53] "CeraLink B58035U* Capacitor for fast-switching semiconductors - Flex Assembly (FA) series." TDK, 2023.

[54] "Design of Planar Power Transformers." Ferroxcube, 1997.

[55] “E32/6/20/R Planar E cores and accessories Datasheet.” Ferroxcube, 2008.

[56] “GB02SLT12-214 1200V 2A SiC Schottky MPS Diode Datasheet.” GeneSiC Semiconductor, 2020.

[57] “GD10MPS12A 1200V 10A SiC Schottky MPS Diode.” GeneSiC Semiconductor, 2021.

[58] “GD10MPS17H 1700V 10A SiC Schottky MPS Diode Datasheet.” GeneSiC Semiconductor, 2021.

[59] “IHLP-5050FD-01 IHLP Commercial Inductors, High Saturation Series Datasheet.” Vishay Dale, 2020.

[60] “A Series PulsEater Ceramic Composition Resistor Datasheet.” Ohmite, 2022.

www.ingramcontent.com/pod-product-compliance
Lightning Source LLC
LaVergne TN
LVHW091038150826
845672LV00006BA/1879